2nd Edition

AP® CALCULUS AB & BC
CRASH COURSE®

Flavia Banu, M.A.
Joan Rosebush, M.A.

Updated by
Stu Schwartz

Research & Education Association
Visit our website at: www.rea.com

Research & Education Association
61 Ethel Road West
Piscataway, New Jersey 08854
E-mail: info@rea.com

AP® CALCULUS AB & BC CRASH COURSE®

Printed in the United States of America

Library of Congress Control Number: 2016944895

ISBN-13: 978-0-7386-1219-5
ISBN-10: 0-7386-1219-7

K16

AP CALCULUS AB & BC CRASH COURSE
TABLE OF CONTENTS

PART I INTRODUCTION

PART II FUNCTIONS, GRAPHS, AND LIMITS

PART III DERIVATIVES

PART IV INTEGRALS

PART V SEQUENCES AND SERIES

PART VI THE EXAM

Online Practice Exam .... *www.rea.com/studycenter*

ABOUT THIS BOOK

REA's *AP Calculus AB & BC Crash Course* is a targeted test prep designed to assist you in your preparation for either version of the AP Calculus exam. This book was developed based on an in-depth analysis of both the latest AP Calculus Course Description outline as well as actual AP test questions.

Written by an AP teacher and a college professor, our easy-to-read format gives students a crash course in Calculus, for both the AB and BC exams.

Unlike other test preps, our *AP Calculus AB & BC Crash Course* gives you a review specifically focused on what you really need to study in order to ace the exam. The review chapters offer you a concise way to learn all the important facts, terms, and concepts before the exam. This revised edition of AP Calculus AB & BC covers all changes to the 2017 exams: L'Hospital's rule will now be included in Calculus AB as well as BC; absolute and conditional convergence of infinite series, the limited comparison test, and the alternating series error have been added to BC topics. In addition, multiple-choice questions will now feature four answer choices instead of five.

The introduction discusses the keys for success and shows you strategies to help you build your overall point score. Parts Two through Five are content review chapters. Here you will find the core of what you need to know for the exam. Part Six shares techniques for using the TI-83 Plus graphing calculator as well as information about the multiple-choice questions and the free-response questions. Our authors show you what you need to know in order to answer correctly the types of questions that will appear on the exams.

No matter how or when you prepare for the AP Calculus AB or BC exams, REA's *Crash Course* will show you how to study efficiently and strategically, so you can get a high score.

To check your test readiness for the AP Calculus AB and BC exams, either before or after studying this *Crash Course*, take REA's **FREE online practice exams** (1 each for AB & BC). To access your practice exam, visit the online REA Study Center at *www.rea.com/studycenter* and follow the on-screen instructions. This true-to-format test features automatic scoring, detailed explanations of all answers, and diagnostic score reporting that will help you identify your strengths and weaknesses so you'll be ready on exam day!

ABOUT OUR AUTHORS

Flavia Banu graduated from Queens College of the City University of New York with a B.A. in Pure Mathematics in 1994, and an M.A. in Pure Mathematics in 1997. From 1994 to 2008, Ms. Banu was an adjunct professor at Queens College, where she taught Algebra and Calculus II. She teaches mathematics at Bayside High School in Bayside, New York, and coaches the school's math team. Her favorite course is AP Calculus because it requires "the most discipline, rigor, and creativity."

Joan Marie Rosebush teaches calculus courses at the University of Vermont. She also serves as the university's Director of Student Success for the College of Engineering and Mathematical Sciences. Ms. Rosebush has taught mathematics to elementary, middle school, high school, and college students. She taught AP Calculus via satellite television to high school students scattered throughout Vermont. Recently, she has been teaching live/online.

Ms. Rosebush earned her bachelor of arts degree in elementary education, with a concentration in mathematics, at the University of New York in Cortland, N.Y. She received her master's degree in education from Saint Michael's College, Colchester, Vermont. She went on to earn a Certificate of Advanced Graduate Study in Administration at Saint Michael's College.

ACKNOWLEDGMENTS

In addition to our authors, we would like to thank Larry B. Kling, Vice President, Editorial, for his overall guidance, which brought this publication to completion; Pam Weston, Publisher, for managing the publication to completion; Diane Goldschmidt, Managing Editor, for project management; and Fred Grayson of American BookWorks Corporation for overseeing manuscript development and typesetting.

We also extend our thanks to Stu Schwartz, for technically editing and reviewing this edition.

PART I
INTRODUCTION

Keys for Success on the AP Calculus AB & BC Exams

Let's face it. The AP Calculus exams are not easy. The course itself is filled with hundreds of formulas, diagrams, and nuances that can be confusing, and by the time you take the actual AP exam, you may be overwhelmed by what you "think" you need to know to get a good score. But don't worry, this *Crash Course* focuses on the key information you *really* need to know for both calculus exams. However, this is not a traditional review book. During the course of the school year, you should learn most of the material that will appear on the AP Calculus exam you're taking. As you go through this book, if you discover that there is something you don't understand, consult your textbook or ask your teacher for clarification.

This *Crash Course* will help you become more pragmatic in your approach to studying for the AP Calculus exams. It's like taking notes on 3 × 5 cards, except we've already done it for you in a streamlined outline format.

 ## I. STRUCTURE OF THE EXAM

Both the AP Calculus AB and BC exams have the same format.

	Number of Questions/ Problems	Time (minutes)
Section I: Multiple Choice Part A: No calculator	30	60
Part B: Graphing calculator allowed	15	45
Section II: Free Response Part A: Graphing calculator allowed	2	30
Part B: No calculator	4	60
TOTAL TIME		195

Each section of the test is worth 50% of your grade. You will have 3¼ hours to complete your exam. According to the College Board, in Section II of Calculus AB and BC, if you complete Part A before your time is up, you *cannot* move on to Part B. So if you finish that section early, you will have time to check your answers. However, if you complete Part B before your 60 minutes are up, you can return to Part A, without the use of the calculator. You will have enough space to work out your problems in the exam booklet.

The multiple-choice answers are scored electronically, and you are not penalized for incorrect answers. Therefore, it makes sense to guess on a question if you don't know the answer.

In the free-response section of the exams, it is important to show your work so the AP readers can evaluate your method of achieving your answers. You will receive partial credit as long as your methods, reasoning, and conclusions are presented in a clear way. You should use complete sentences when answering the questions in this portion of the exam.

For those questions requiring the use of a graphing calculator, the scorers will want to see your mathematical setup that led to the solution provided by the calculator. You should demonstrate the equation being solved, derivatives being evaluated, and so on. Your answers should be in standard mathematical notation.

If a calculation is given as a decimal approximation, it should be correct to three places following the decimal point, unless you are asked for something different in the question.

Changes to the AP Calculus Exams

Starting with the administration of the 2017 AP Calculus exams, there will be several changes to the curriculum: L'Hôspital's rule will now be a part of the AB exam. It is unknown to what extent

L'Hôspital's rule will be covered in the AB exam. It is likely that the simpler use of L'Hôspital's rule using the $\frac{0}{0}$ and $\frac{\infty}{\infty}$ forms will be covered in the AB exam, while the more complicated indeterminate forms ($\infty - \infty$, $0 \cdot \infty$, 1^∞, 0^0, ∞^0) will be part of the BC exam. Absolute and conditional convergence of infinite series has been added to the BC exam topics, as well as the limit comparison test and the alternating series error. All of these topics are covered in this book.

Also, please note that although Newton's method, cylindrical shells, and powers of trig functions are no longer tested on the AP Calculus exams, we have included the information in this book because of its overall usefulness.

The biggest change is that the multiple-choice questions in the AB exam will have four choices rather than five. All multiple-choice sample problems in this book and in the online practice tests incorporate this change.

 ## II. THE SCORES

The scores from Section I and Section II are combined to create a composite score.

AP SCORE SCALE

5	Extremely well qualified
4	Well qualified
3	Qualified
2	Possibly qualified
1	No recommendation

To be "qualified" is to receive college credit or advanced placement. However, the acceptance of these scores for credit is at the discretion of the individual college. You should check with the colleges to which you are applying to see what AP scores they accept for college credit or advanced placement.

According to the statistics reported by the College Board, 57.7% of all the students who took the AP Calculus AB exam in 2013 scored a 3 or higher and 22.6% received a 5. Almost 46% scored a 5 on the AP Calculus BC exam.

Remember, if you're taking the AP Calculus BC exam, you will also receive a Calculus AB subscore for that part of the Calculus BC exam that covers AB topics.

What does this all mean? Why do so many more students who take the AP Calculus BC exam score a 5? It has to do with the level of the student taking the BC exam. It doesn't mean they're smarter, but rather that they've already been through the Calculus AB course, which represents about 40% of the BC-level exam. Thus, it stands to reason that the BC exam would, in effect, be much easier. After all, almost 85% who took the BC exam scored a 3 or higher on the AB portion of the BC test.

Most of this shouldn't have too much impact on how you do on the exam. If you study for the test using this *Crash Course* book and pay attention during the school year, you will likely be pleasantly surprised when you receive your scores.

 ## III. STRATEGIES FOR SCORING HIGH

Keep in mind that one of the best ways to prepare for your exam is to research past exams. Although the AP Calculus exams have changed slightly for 2017, it still makes sense to go back to previous tests and answer the questions. The single most important aspect of scoring high on any standardized test is to have complete familiarity with the questions that will be asked on the test. You may not find the exact questions, but you will find those that are similar in content to questions you will encounter on your exam.

On the College Board website you will find past free-response questions posted. The more questions you answer in preparation for the test, the better you will do on the actual exam.

On the actual exam, make sure you write clearly. This sounds like a very simple thing, but if the AP Reader scoring your exam cannot read your answer, you will lose credit. We suggest that you cross out your incorrect work rather than erase it.

Along those lines, keep in mind that because you will be graded on your method of calculations, make sure you show all of your work. Clearly identify functions, graphs, tables, or any other items that you've included in order to reach your conclusions.

Read the graphs carefully, as well as the questions. Make sure they correspond and that you are dealing with like terms.

You do not need to simplify numeric or algebraic answers. Decimal approximations should be correct to three places—unless stated otherwise.

IV. USING SUPPLEMENTAL INFORMATION

This *Crash Course* contains everything you need to know in order to score well on either the AP Calculus AB or BC exams. The AP Calculus Course Description Booklet published by the College Board can also be a very useful tool in your studies (*www.collegeboard.org*).

In addition, REA's *AP Calculus AB & BC All Access*® Book + Web + Mobile study system further enhances your exam preparation by offering a comprehensive review book plus a suite of online assessments (topic-level quizzes, mini-tests, full-length practice tests, and e-flashcards), all designed to pinpoint your strengths and weaknesses and help focus your study for the exam.

Good luck on your AP Calculus exams!

PART II
FUNCTIONS, GRAPHS, AND LIMITS

Analysis of Graphs

I. ANALYSIS OF GRAPHS

A. Basic Functions—you need to know how to graph the following functions and any of their transformations by hand.

1. Polynomials, absolute value, square root functions

i. $y = x$	ii. $y = x^2$	iii. $y = x^3$
Linear function	Quadratic function	Cubic function
Domain: $(-\infty, \infty)$	Concave up	Increasing
Range: $(-\infty, \infty)$	Min. point $(0,0)$	Domain: $(-\infty, \infty)$
Odd	Domain: $(-\infty, \infty)$	Range: $(-\infty, \infty)$
	Range: $[0, \infty)$	No min/max
	Even	Odd

iv. $y = |x| = \begin{cases} x, & x \geq 0 \\ -x, & x < 0 \end{cases}$ Absolute value function V-shape opens upward Not differentiable at $x = 0$ Domain: $(-\infty, \infty)$ Range: $[0, \infty)$ Even	v. $y = \sqrt{x}$ Square root function Strictly increasing Not differentiable at $x = 0$ Domain: $[0, \infty)$ Range: $[0, \infty)$
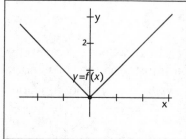	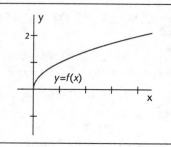

2. Trigonometric functions

i. $y = \sin(x)$ x-intercepts: $x = n\pi$, n is an integer y-intercept: $y = 0$ Odd	ii. $y = \cos(x)$ x-intercepts: $x = (2n+1)\dfrac{\pi}{2}$, n is an integer y-intercept: $y = 1$ Even
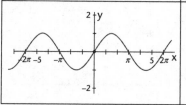	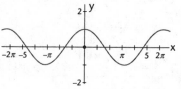

iii. $y = \tan(x)$
x-intercepts: $x = n\pi$, n is an integer
vertical asymptotes: $x = (2n+1)\dfrac{\pi}{2}$, n is an integer
y-intercept: $y = 0$
Odd

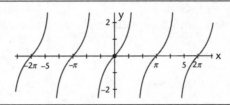

3. Inverse trigonometric functions and their domain and range

i. $y = \sin^{-1}(x)$	ii. $y = \cos^{-1}(x)$
Domain: [−1, 1] Range: $\left[-\dfrac{\pi}{2}, \dfrac{\pi}{2}\right]$ Strictly increasing	Domain: [−1, 1] Range: $[0, \pi]$ Strictly decreasing
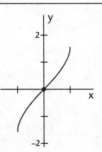	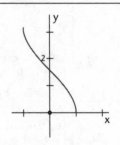

iii. $y = \tan^{-1}(x)$
Domain: $(-\infty, \infty)$ Range: $\left(-\dfrac{\pi}{2}, \dfrac{\pi}{2}\right)$

Horizontal Asymptotes: $y = \pm\dfrac{\pi}{2}$, strictly increasing

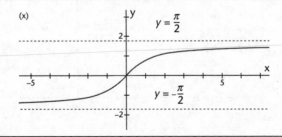

4. Exponential and Natural Logarithmic functions

i. $y = e^x$	ii. $y = \ln(x)$
This is the inverse of $y = \ln(x)$	This is the inverse of $y = e^x$
Strictly increasing	Strictly increasing
x-intercepts: none	x-intercept: $x = 1$
y-intercept: $y = 1$	y-intercepts: none
horizontal asymptote: $y = 0$	vertical asymptote: $x = 0$
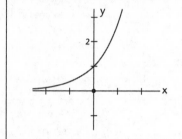	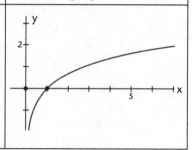

5. Rational functions

i. $y = \dfrac{1}{x}$	ii. $y = \dfrac{1}{x^2}$
Undefined at $x = 0$	Undefined at $x = 0$
Not differentiable at $x = 0$	Not differentiable at $x = 0$
x-intercepts: none	x-intercepts: none
y-intercepts: none	y-intercepts: none
horizontal asymptote: $y = 0$	horizontal asymptote: $y = 0$
vertical asymptote: $x = 0$	vertical asymptote: $x = 0$
$\lim\limits_{x \to 0} f(x)$ *dne* (does not exist)	$\lim\limits_{x \to 0} f(x) = \infty$
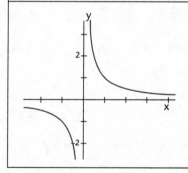	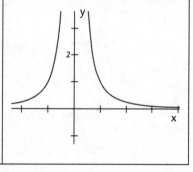

6. Piecewise functions

i. $y = \begin{cases} x, & x \le 0 \\ x^2, & x > 0 \end{cases}$

Continuous at $x = 0$ (because the y value of each piece at $x = 0$ is the same, $y = 0$)
Not differentiable at $x = 0$ (because the derivatives of the pieces at $x = 0$ are not equal)

ii. $y = \begin{cases} x+1, & x \le 0 \\ x^2, & x > 0 \end{cases}$

Discontinuous at $x = 0$ (because here the y value of the left side of the graph is 1 ($0 + 1 = 1$) and the y value of the right side of the graph does not exist.)
Not differentiable at $x = 0$ (since it's not continuous there)

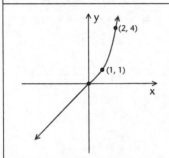

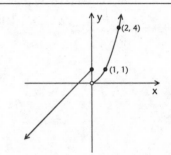

iii. $y = \begin{cases} x^3, & x \le 0 \\ x^2, & x > 0 \end{cases}$

Continuous at $x = 0$ (because the y value of each piece at $x = 0$ is the same, $y = 0$)
Differentiable at $x = 0$ (because the two pieces have equal derivatives at $x = 0$)

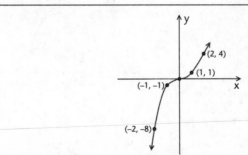

7. Circle Equations

   i. Upper semicircle with radius a and center at the origin: $y = \sqrt{a^2 - x^2}$. This is a function. For example, $y = \sqrt{9 - x^2}$

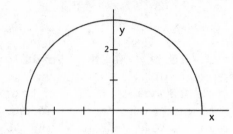

   ii. Lower semicircle with radius a and center at the origin: $y = -\sqrt{a^2 - x^2}$. This is a function. For example, $y = -\sqrt{9 - x^2}$

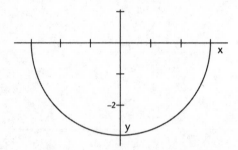

   iii. Circle with radius a and center at the origin: $x^2 + y^2 = a^2$. This is not a function since some x-values correspond to more than one y-value. For example, $x^2 + y^2 = 9$

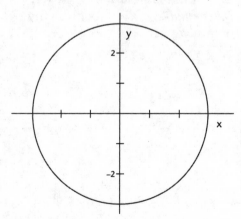

iv. Circle with radius a and center at (b, c): $(x - b)^2 + (y - c)^2 = a^2$. This is not a function either. For example, $(x - 2)^2 + (y + 3)^2 = 9$

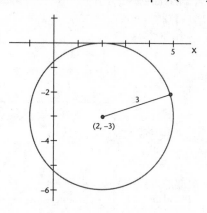

8. **Summary of Basic Transformations of Functions**

   A. Making changes to the equation of $y = f(x)$ will result in changes in its graph. The following transformations occur most often.

Transformation	$g(x)$ is obtained when	Example
$y = -f(x)$	$y = f(x)$ reflects in the x-axis	$f(x) = x^2$ $g(x) = -x^2$

$y = f(x + a),$ $a > 0$	$y = f(x)$ translates a units left (horizontal shift)	
$y = f(x - a),$ $a > 0$	$y = f(x)$ translates a units right (horizontal shift)	
$y = f(x) + a,$ $a > 0$	$y = f(x)$ translates a units up (vertical shift)	
$y = f(x) - a,$ $a > 0$	$y = f(x)$ translates a units down (vertical shift)	

$y = af(x)$ $0 \le a < 1$	$y = f(x)$ widens	
$y = af(x)$ $a > 1$	$y = f(x)$ narrows	

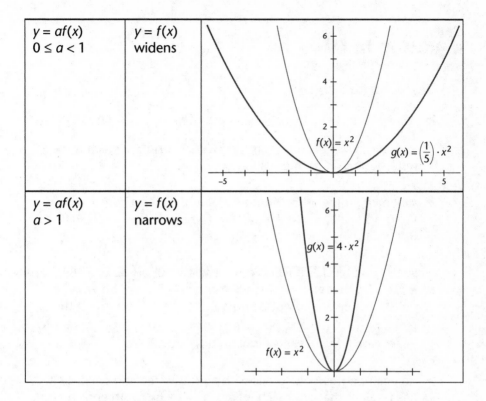

B. For trigonometric functions, $f(x) = a \sin(bx + c) + d$ or $f(x) = a \cos(bx + c) + d$, a is the amplitude (half the height of the function), b is the frequency (the number of times that a full cycle occurs in a domain interval of 2π units), $\frac{c}{b}$ is the horizontal shift and d is the vertical shift.

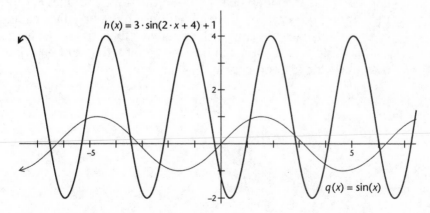

Keep in Mind...

➤ $\dfrac{1}{\sin(x)} \neq \sin^{-1}(x)$. The reciprocal of $\sin(x)$, $\dfrac{1}{\sin(x)}$, is equivalent
to $\csc(x)$, whereas $\sin^{-1}(x)$ is the inverse of $\sin(x)$, which is the reflection of $\sin(x)$ in the line $y = x$.

➤ When changing a function by adding a positive constant to x, the graph will shift to the *left*, not the right. The graph shifts to the right a units when a is subtracted from x.

➤ When graphing a function on the calculator (TI-83 or TI-84), make sure that all the plots are turned off; otherwise you risk getting an error and not being able to graph. To turn off the plots, press Ⓨ ⚌ and place the cursor on the plot you want to deactivate (whichever is highlighted). Press Enter.

➤ An even-degree polynomial with a positive leading coefficient has y-values that approach infinity as $x \rightarrow \pm\infty$ (both ends go up). If the polynomial has a negative leading coefficient, its y-values approach negative infinity as $x \rightarrow \pm\infty$ (both ends go down).

➤ An odd-degree polynomial with a positive leading coefficient has y-values that approach infinity as $x \rightarrow \infty$ and y-values that approach negative infinity as $x \rightarrow -\infty$ (the right end goes up and the left end goes down). If the polynomial has a negative leading coefficient its y-values approach negative infinity as $x \rightarrow \infty$; as $x \rightarrow -\infty$ its y-values approach positive infinity (the right end goes down and the left end goes up).

CHAPTER 2
PRACTICE PROBLEMS
(See solutions on page 195)

For each of the functions in problems 1–8, draw the mother function and the given function on the same set of axes.

1. $y = \dfrac{1}{2}(x+1)^3 - 3$

2. $y = 2|3x+4|$

3. $y = \sqrt{x-6} + 1$

4. $y = -\dfrac{3}{x} + 2$

5. $y = e^{x+2} - 1$

6. $y = \ln(4-x)$

Limits of Functions

I. MEANING OF LIMIT

A. The limit of a function, $y = f(x)$, as x approaches a number or $\pm\infty$, represents the value that y <u>approaches</u>.

B. The *left-hand* limit, $\lim\limits_{x \to a^-} f(x) = L$, states that as x approaches a, <u>from the *left* of a</u>, $f(x)$ approaches L.
The *right-hand* limit, $\lim\limits_{x \to a^+} f(x) = L$, states that as x approaches a, <u>from the *right* of a</u>, $f(x)$ approaches L.

C. The expression $\lim\limits_{x \to a} f(x) = L$ states that as x approaches a, <u>simultaneously from the *left and right* of a</u>, $f(x)$ approaches L.

D. The limit of a function at a point exists if and only if the left- and right-hand limits exist *and* are equal.
Symbolically, if $\lim\limits_{x \to a^-} f(x) = L$ and $\lim\limits_{x \to a^+} f(x) = L$ then $\lim\limits_{x \to a} f(x) = L$.
The converse is also true.

E. If the left- and right-hand limits are not equal at a given x value then the limit at the given x value does not exist.
Symbolically, if $\lim\limits_{x \to a^-} f(x) \neq \lim\limits_{x \to a^+} f(x)$ then $\lim\limits_{x \to a} f(x)$ does not exist.
The converse is also true.

II. EVALUATING LIMITS ALGEBRAICALLY

A. Generally, $\lim\limits_{x \to a} f(x) = f(a)$. That is, to evaluate the limit of a function algebraically, substitute x with the value x approaches.

(If $x \to \infty$ or $x \to -\infty$, substitute x with values that are very large or very small, respectively.)

1. If $\lim\limits_{x \to a} f(x) = \dfrac{b}{0}$, $b \neq 0$, then take the left- and right-hand limits separately to see if they are the same or not. (In this case, $x = a$ is a *vertical asymptote* of $y = f(x)$.) For instance, $\lim\limits_{x \to 0} \dfrac{1}{x} = \dfrac{1}{0}$ after substituting 0 for x. Since the left-hand limit, $\lim\limits_{x \to 0^-} \dfrac{1}{x} = -\infty$, and the right-hand limit, $\lim\limits_{x \to 0^+} \dfrac{1}{x} = \infty$ are not the same, $\lim\limits_{x \to 0} \dfrac{1}{x}$ does not exist. Similarly, $\lim\limits_{x \to 0} \dfrac{1}{x^2} = \dfrac{1}{0}$. However, the left-hand limit, $\lim\limits_{x \to 0^-} \dfrac{1}{x^2} = \infty$, and the right-hand limit, $\lim\limits_{x \to 0^+} \dfrac{1}{x^2} = \infty$, so $\lim\limits_{x \to 0} \dfrac{1}{x^2} = \infty$.

2. Indeterminate forms:

$\lim\limits_{x \to a} f(x)$ or $\lim\limits_{x \to \pm\infty} f(x)$	Method
$\dfrac{0}{0}$	L'Hôspital's Rule (see chapter 7) or simplify by factoring first. For example, using L'Hôspital's Rule, $\lim\limits_{x \to 2} \dfrac{x-2}{x^2-4} = \lim\limits_{x \to 2} \dfrac{1}{2x} = \dfrac{1}{4}$ or, by factoring, $\lim\limits_{x \to 2} \dfrac{x-2}{x^2-4} = \lim\limits_{x \to 2} \dfrac{1}{x+2} = \dfrac{1}{4}$.
$\dfrac{\infty}{\infty}$	L'Hôspital's rule or if $x \to \infty$ or $x \to -\infty$ consider only the leading terms when taking the limit since the rest are negligible. For example, by L'Hôspital's Rule, $\lim\limits_{x \to \infty} \dfrac{x^2-3x+2}{x^2+2x-4} = \lim\limits_{x \to \infty} \dfrac{2x}{2x} = 1$ or, by considering only the leading terms, $\lim\limits_{x \to \infty} \dfrac{x^2-3x+2}{x^2+2x-4} = \lim\limits_{x \to \infty} \dfrac{x^2}{x^2} = 1$.

$\lim\limits_{x \to a} f(x)$ *or* $\lim\limits_{x \to \pm\infty} f(x)$	Method
$0 \bullet \infty$	Algebraically manipulate to change expression to $\dfrac{0}{0}$ or $\dfrac{\infty}{\infty}$ and then use L'Hôspital's rule. For example, $$\lim_{x \to \infty} \frac{1}{x} e^x = \lim_{x \to \infty} \frac{e^x}{x} = \lim_{x \to \infty} \frac{e^x}{1} = \infty.$$
$\infty - \infty$	Combine the two expressions into one and then take the limit. For example, in this case, rationalize the numerator: $$\lim_{x \to \infty} \sqrt{x^2 + x} - x =$$ $$\lim_{x \to \infty} \sqrt{x^2 + x} - x \cdot \frac{\sqrt{x^2 + x} + x}{\sqrt{x^2 + x} + x}$$ $$= \lim_{x \to \infty} \frac{x}{\sqrt{x^2 + x} + x}$$ $$= \lim_{x \to \infty} \frac{x}{2x} = \frac{1}{2}.$$
$1^\infty,\ \infty^0,\ 0^0$	Let $y = f(x)$. Then evaluate $\lim\limits_{x \to a} \ln(y)$ using one of the above methods. The final answer is $e^{\lim\limits_{x \to a} \ln(y)}$. An example of a 1^∞ case: $\lim\limits_{x \to 0} (1 + x)^{\frac{1}{x}}$. Let $y = (1 + x)^{\frac{1}{x}}$, so $\ln y = \ln(1 + x)^{\frac{1}{x}} = \frac{1}{x} \ln(1 + x) = \frac{\ln(1 + x)}{x}$. Now, $\lim\limits_{x \to 0}(\ln y) = \lim\limits_{x \to 0} \frac{\ln(1 + x)}{x} = \lim\limits_{x \to 0} \frac{1}{1 + x} = 1$, by L'Hôspital's rule. So, $\lim\limits_{x \to 0}(\ln y) = 1$ implies that $\lim\limits_{x \to 0}(y) = e^1 = e$.

Test Tip

Use of L'Hôspital's Rule is new to the AB exam and it is unknown to what extent it will be covered. Most likely, the simpler forms of $\dfrac{0}{0}$ and $\dfrac{\infty}{\infty}$ will be covered in the AB exam while the more complicated indeterminate cases ($\infty - \infty$, $0 \cdot \infty$, 1^∞, 0^0, ∞^0) will remain as part of the BC exam.

III. EVALUATING LIMITS GRAPHICALLY

A. Common limit concepts.

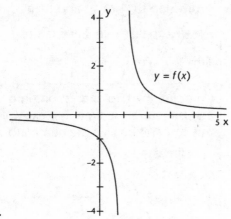

1.

$$\lim_{x \to 1} f(x) = \textit{dne} \text{ because the left-hand limit is } -\infty \text{ and the}$$

right-hand limit is ∞.

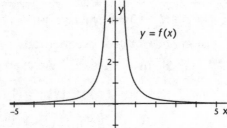

2.

$$\lim_{x \to 0} f(x) = \infty \text{ because the left- and right-hand limits are both } \infty.$$

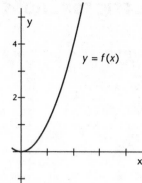

3.

$\lim\limits_{x\to 1} f(x) = 1$ and $f(1) = 1$. The function value and the limit at $x = 1$ are equal.

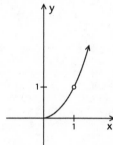

4.

$\lim\limits_{x\to 1} f(x) = 1$ and $f(1)$ *dne*. The function is undefined at $x = 1$ but the limit exists there.

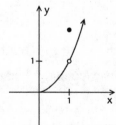

5.

$\lim\limits_{x\to 1} f(x) = 1$ and $f(1) = 2$. The function value is not equal to the limit at $x = 1$.

IV. LIMITS INVOLVING TRIGONOMETRIC FUNCTIONS

A. $\lim\limits_{x\to 0}\dfrac{\sin(x)}{x}=1$ and $\lim\limits_{x\to 0}\dfrac{\cos(x)-1}{x}=0$ are the most common trigonometric limits; also:

1. $\lim\limits_{x\to 0}\dfrac{\sin(ax)}{ax}=1$

2. $\lim\limits_{x\to 0}\dfrac{\sin(ax)}{bx}=\dfrac{a}{b}$

3. $\lim\limits_{x\to 0}\dfrac{x}{\sin(x)}=1$

4. $\lim\limits_{x\to 0}\dfrac{ax}{\sin(ax)}=1$

5. $\lim\limits_{x\to 0}\dfrac{ax}{\sin(bx)}=\dfrac{a}{b}$

6. $\lim\limits_{x\to 0}\dfrac{\sin ax}{\sin bx}=\dfrac{a}{b}$

7. $\lim\limits_{x\to 0}\dfrac{\tan ax}{bx}=\dfrac{a}{b}$

8. $\lim\limits_{x\to 0}\dfrac{\tan ax}{\tan bx}=\dfrac{a}{b}$

9. $\lim\limits_{x\to 0}\dfrac{\sin ax}{\tan bx}=\dfrac{a}{b}$

10. $\lim\limits_{x\to 0}\dfrac{\tan(ax)}{\sin(bx)}=\dfrac{a}{b}$

All the above limits can be found using L'Hôspital's rule as well. However, knowing the answer without doing the work will save you valuable time!

V. LIMITS INVOLVING e.*

A. Basic definitions of e: $\lim\limits_{x\to\infty}\left(1+\dfrac{1}{x}\right)^{x}=e$ or $\lim\limits_{x\to 0}(1+x)^{\frac{1}{x}}=e$

1. The above definitions can be used to evaluate the following common limits (with minimal algebraic manipulation):

i. $\lim\limits_{x\to\infty}\left(1+\dfrac{1}{x}\right)^{2x}=\lim\limits_{x\to\infty}\left[\left(1+\dfrac{1}{x}\right)^{x}\right]^{2}=[e]^{2}=e^{2}$

ii. $\lim\limits_{x\to\infty}\left(1+\dfrac{1}{3x}\right)^{x}=\lim\limits_{x\to\infty}\left[\left(1+\dfrac{1}{3x}\right)^{3x}\right]^{\frac{1}{3}}=[e]^{\frac{1}{3}}=e^{\frac{1}{3}}$

iii. $\lim\limits_{x\to\infty}\left(1-\dfrac{1}{x}\right)^{x}=\lim\limits_{x\to\infty}\left[\left(1+\dfrac{1}{-x}\right)^{-x}\right]^{-1}=e^{-1}$

iv. $\lim\limits_{x\to\infty}\left(1+\dfrac{3}{x}\right)^{x}=\lim\limits_{x\to\infty}\left[\left(1+\dfrac{1}{\frac{1}{3}x}\right)^{\frac{1}{3}x}\right]^{3}=e^{3}$

v. $\lim\limits_{x\to\infty}\left(2+\dfrac{1}{x}\right)^{x}=\lim\limits_{x\to\infty}\left[2\left(1+\dfrac{1}{2x}\right)\right]^{x}=\lim\limits_{x\to\infty}\left[2^{x}\left(1+\dfrac{1}{2x}\right)^{x}\right]=$

$\lim\limits_{x\to\infty}2^{x}\lim\limits_{x\to\infty}\left(1+\dfrac{1}{2x}\right)^{x}=\lim\limits_{x\to\infty}2^{x}\lim\limits_{x\to\infty}\left[\left(1+\dfrac{1}{2x}\right)^{2x}\right]^{\frac{1}{2}}=\infty\cdot e^{\frac{1}{2}}=\infty$

vi. $\lim\limits_{x\to 0}(1+3x)^{\frac{1}{x}}=\lim\limits_{x\to 0}\left[(1+3x)^{\frac{1}{3x}}\right]^{3}=e^{3}$

These limits can all be evaluated using natural logs but the definitions of e, along with the algebraic manipulations, save you a lot of time when you consider that you have less than 2 minutes for each multiple-choice question.

*This is a BC Calculus topic.

Keep in Mind...

➤ The answer to a limit question can only be one of the following: a number, $-\infty$, ∞, or "does not exist."

➤ Try to imagine the graph of a function when taking the function's limit.

➤ The limit of a function as x approaches a may or may not be the same as the value of the function at $x = a$.

➤ If a function's limit exists, it must equal a number. A function's limit does not exist in two cases: when the limit from the left and right of the x-value are unequal, or when the y-values approach $\pm\infty$.

➤ Beware the indeterminate forms!

CHAPTER 3
PRACTICE PROBLEMS
(See solutions on page 197)

1. $\lim\limits_{x \to \infty} \dfrac{1 - 3x + 6x^2 - x^{10}}{2 + 4x^4 - 8x^7 + 8x^{10}} =$

    (A) 0

    (B) ∞

    (C) $-\infty$

    (D) $-\dfrac{1}{8}$

2. $\lim\limits_{x \to \infty} \dfrac{\sin x}{x} =$

(A) 0

(B) ∞

(C) 1

(D) Does not exist

3. $\lim\limits_{x \to 0^+} \ln(x) =$

(A) 1

(B) -1

(C) 0

(D) $-\infty$

4. $\lim\limits_{x \to 3} \dfrac{\sqrt{x+6} - 3}{x - 3} =$

(A) 1

(B) -1

(C) $\dfrac{1}{6}$

(D) ∞

5. $\lim\limits_{x \to 1} \dfrac{3}{x - 1} =$

(A) 1

(B) ∞

(C) $-\infty$

(D) Does not exist

Asymptotes and Unbounded Behavior

I. ASYMPTOTES

A. Asymptotes are vertical or horizontal lines (the AP Calculus exams do not include oblique asymptotes) that a graph approaches. Polynomials do not have asymptotes. The functions that most commonly have asymptotes are rational functions, $y = \dfrac{f(x)}{g(x)}$ as well as other functions such as $y = e^x$, $y = \ln(x)$, $y = \tan(x)$, $y = \cot(x)$, $y = \sec(x)$, $y = \csc(x)$, and their transformations.

1. Vertical asymptotes are vertical lines that a graph only approaches but *never intersects*. (Well, almost never! See a rare example of an exception below.)

   i. $x = k$ is a vertical asymptote of $y = f(x)$ if and only if
   $$\lim_{x \to k^-} f(x) = \pm\infty, \text{ or } \lim_{x \to k^+} f(x) = \pm\infty, \text{ or both.}$$

   ii. For a rational function $y = \dfrac{f(x)}{g(x)}$, the equation of a vertical asymptote is $x = k$ if and only if $g(k) = 0$ and $f(k) \neq 0$; if, however, $f(a) = g(a) = 0$, then, in most cases, there is a removable discontinuity (a hole, not a vertical asymptote) at $x = a$. An example of an exception to the rule is $y = \dfrac{x}{|x|}$ which has an irremovable (nonremovable) discontinuity at $x = 0$ but no vertical asymptote. Also, the function $y = \dfrac{x-1}{x^2-1}$ has one hole (at $x = 1$, where both numerator and denominator are 0) and one vertical asymptote (at $x = -1$, where only the denominator is 0). See graphs below:

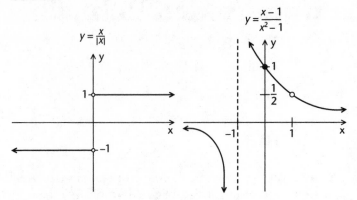

iii. A function can have an infinite number of vertical asymptotes (for example, $y = \tan(x)$)

Rare example of a graph intersecting its vertical asymptote:

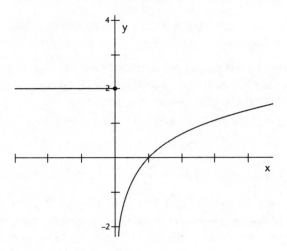

$$f(x) = \begin{cases} 2, & x \le 0 \\ \ln(x), & x > 0 \end{cases}$$

The vertical asymptote for $f(x)$ is $x = 0$ and $f(0) = 2$, thus the point $(0,2)$ is on the vertical asymptote.

2. Horizontal asymptotes are horizontal lines that a graph approaches and *may intersect*.

i. $y = k$ is a horizontal asymptote for $y = f(x)$ if and only if $\lim\limits_{x\to\infty} f(x) = k$, or $\lim\limits_{x\to-\infty} f(x) = k$, or both. Horizontal asymptotes give us an idea of the function's end behavior (as $x \to \pm\infty$). The graph of $y = \dfrac{x}{e^{x^2}}$, below, has only one horizontal asymptote, $y = 0$, since $\lim\limits_{x\to\infty} \dfrac{x}{e^{x^2}} = 0$ and $\lim\limits_{x\to-\infty} \dfrac{x}{e^{x^2}} = 0$. Note that the graph intersects its asymptote at $(0,0)$.

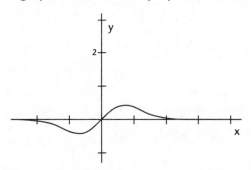

ii. A function can have at most two horizontal asymptotes. The function $y = \tan^{-1}(x)$, below, has two horizontal asymptotes, $y = \pm\dfrac{\pi}{2}$ since $\lim\limits_{x\to\infty} \tan^{-1}(x) = \dfrac{\pi}{2}$ and $\lim\limits_{x\to-\infty} \tan^{-1}(x) = -\dfrac{\pi}{2}$.

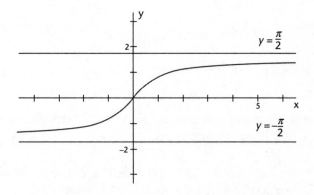

iii. A function which only has one horizontal asymptote does not have to approach this asymptote on both ends! The function $y = \dfrac{x}{e^x}$, below, has only one horizontal asymptote,

$y = 0$ since $\lim\limits_{x \to \infty} \dfrac{x}{e^x} = 0$ and $\lim\limits_{x \to -\infty} \dfrac{x}{e^x} = -\infty$ and so it only approaches this asymptote as $x \to \infty$.

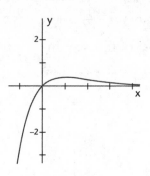

II. UNBOUNDED BEHAVIOR

A. If a function, $y = f(x)$, approaches positive infinity either as $x \to a$ or as $x \to \pm\infty$, the function is said to increase without bound. Similarly, if a function, $y = f(x)$, approaches negative infinity either as $x \to a$ or as $x \to \pm\infty$, the function is said to decrease without bound.

 **Keep in Mind...**

➤ Remember that a graph *might* cross both its horizontal, as well as its vertical, asymptotes!

➤ Do not confuse vertical/horizontal asymptotes with vertical/horizontal tangent lines.

➤ When finding a vertical asymptote, find the root of the denominator and then make sure that it is not also a root of the numerator. To be safe, simplify a rational function completely before finding its vertical asymptotes.

➤ Not only rational functions have asymptotes. Functions such as $y = \ln(x)$, $y = e^x$, $y = \tan(x)$ and their transformations have asymptotes as well.

CHAPTER 4
PRACTICE PROBLEMS

(See solutions on page 198)

Find the horizontal and vertical asymptotes of the following functions:

1. $f(x) = \dfrac{3x^2 - 9x}{x^2 - 9}$

2. $f(x) = \dfrac{x^3 + 3x^2 - 1}{4 - x^3}$

3. What conclusion can you draw about the asymptotes of $f(x)$ if:

(A) $\lim\limits_{x \to \infty} f(x) = 7$

(B) $\lim\limits_{x \to -\infty} f(x) = -\infty$

(C) $\lim\limits_{x \to 4} f(x) = \infty$

Continuity as a Property of Functions

I. CONTINUITY OF FUNCTIONS

A. A function is either continuous (no breaks whatsoever) or discontinuous at certain points.

1. A function $y = f(x)$ is continuous at a point, $x = a$, if and only if $\lim_{x \to a^-} f(x) = \lim_{x \to a^+} f(x) = f(a)$. Simply put, this states that for $y = f(x)$ to be continuous at $x = a$, the limits of the function from the left and right of a must be equal to each other and also equal to the value of the function at $x = a$.

2. All polynomials are continuous.

3. Some of the most common discontinuous functions come in the form of rational functions, piecewise functions and $y = \tan(x)$, $y = \cot(x)$, $y = \sec(x)$, $y = \csc(x)$ and their transformations.

4. A *removable discontinuity* occurs when an otherwise continuous graph has a point (or more) missing. That is, $y = f(x)$ has a removable discontinuity at $x = a$ if and only if $\lim_{x \to a^-} f(x) = \lim_{x \to a^+} f(x) = L$ but $f(a) \neq L$ or $f(a)$ does not exist.

   i. The function $f(x) = \begin{cases} x, & x \neq 1 \\ 2, & x = 1 \end{cases}$ has a removable discontinuity at $x = 1$ since $\lim_{x \to 1^-} f(x) = \lim_{x \to 1^+} f(x) = 1$ and $f(1) \neq 1$. In this case, $f(1) = 2$.

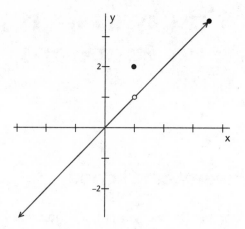

ii. The function $y = \dfrac{\sin(x)}{x}$ has a removable discontinuity at

$x = 0$ since $\displaystyle\lim_{x \to 0^-} \frac{\sin(x)}{x} = \lim_{x \to 0^+} \frac{\sin(x)}{x} = 1$ but $f(0) \neq 1$. In this

case, $f(0)$ does not exist.

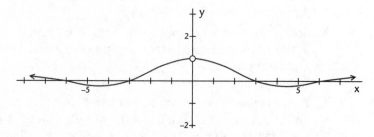

iii. A removable discontinuity is one that can be "filled in" (or removed) if the function is appropriately redefined. To remove the discontinuity of the function in part *i*, the function can be redefined as $f(x) = x$. To remove the discontinuity in part *ii*. The function can be redefined as

$$y = \begin{cases} \dfrac{\sin(x)}{x} & , x \neq 0 \\ 1 & , x = 0 \end{cases}$$

5. A *nonremovable discontinuity* occurs at step breaks in the graph or at vertical asymptotes. That is, $y = f(x)$ has a nonremovable discontinuity at $x = a$ if and only if $\displaystyle\lim_{x \to a^-} f(x) \neq \lim_{x \to a^+} f(x)$ or if one or both of these limits is $\pm\infty$.

i. The function $y = \dfrac{1}{x}$ has a nonremovable discontinuity at $x = 0$ (also a vertical asymptote there) because the function cannot be redefined so that it will be continuous there. This is also called an infinite discontinuity.

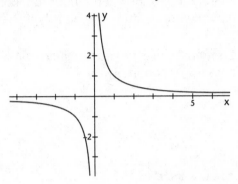

ii. The function $y = \dfrac{x}{|x|}$ has a nonremovable discontinuity at $x = 0$ (a step break, not a vertical asymptote) because we cannot redefine it so that it will become continuous there. This is also called a jump discontinuity.

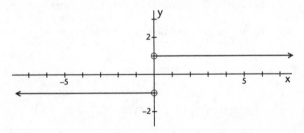

6. If a function is continuous, it does not have to be differentiable! (A function is continuous at a cusp or corner, yet it is not differentiable there; more on differentiability later.)

Keep in Mind...

➤ Loosely speaking, a continuous function is one which can be drawn without lifting the pencil off the paper.

➤ Continuity does not imply differentiability. Differentiability does imply continuity.

CHAPTER 5
PRACTICE PROBLEMS

(See solutions on page 199)

1. Find the x-values for which $f(x) = \dfrac{2}{\sqrt{1-x}}$ is continuous.

2. Find the discontinuities of $f(x) = \dfrac{x^2 + 5x + 6}{x^2 - 4}$ and categorize them as removable or nonremovable.

3. Find all x-values for which $f(x) = \begin{cases} 2 - x, & x < -1 \\ \dfrac{1}{x}, & -1 \le x \le 2 \\ \dfrac{1}{2}, & x > 2 \end{cases}$ is discontinuous.

Parametric, Polar, and Vector Equations*

I. PARAMETRIC AND VECTOR EQUATIONS

A. Parametric and vector equations are used to describe the motion of a body. They have different notations but describe the same concept. An additional variable is involved, called the parameter, usually denoted by t (for time). The parameter does not appear on the graph; it only represents the time at which a given particle is at a given point. Both parametric and vector equations are represented on the Cartesian coordinate system.

1. Parametric equations create only one graph though they contain two equations. They are denoted by: $\begin{cases} x(t) = f(t) \\ y(t) = g(t) \end{cases}$.

2. A vector equation is denoted by: $\langle x(t), y(t) \rangle = \langle f(t), g(t) \rangle$ or, $r(t) = (f(t))i + (g(t))j$.

3. A parametric equation can be written in Cartesian form (x–y form) by using algebraic manipulation.

    i. For example, parametric equations: $\begin{cases} x(t) = t \\ y(t) = t^2 \end{cases}$, $t \geq 0$, can be written in vector form as $\langle x(t), y(t) \rangle = \langle t, t^2 \rangle$ or $r(t) = (t)i + (t^2)j$ and in Cartesian form as $y = x^2$. See graph at right.

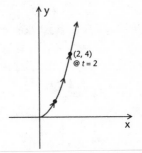

(2, 4) @ $t = 2$

*These are BC Calculus topics.

ii. The difference between a parametric/vector curve and a Cartesian curve is that a parametric/vector curve has *direction*. When drawing a parametric/vector curve, in part 3i, the direction must be specified with arrows, on the graph, in the direction of increasing parameter, or else the graph will be considered incomplete.

II. POLAR EQUATIONS

A. A polar equation, $r = f(\theta)$, is written using polar coordinates (r, θ), where r represents the point's distance from the origin, and θ represents the measurement of the angle between the positive x-axis and the line segment between the point and the origin. Angle θ is measured counterclockwise from the positive x-axis.

B. To switch from Cartesian form to polar form use: $r^2 = x^2 + y^2$ and $\theta = \tan^{-1}\left(\dfrac{y}{x}\right)$; to switch from polar form to Cartesian form, use $x = r\cos(\theta)$ and $y = r\sin(\theta)$.

i. To change $x^2 + y^2 = 9$ from Cartesian form to polar form, rewrite $x^2 + y^2 = 9$ as $r^2 = 9 \to r = 3$ (or $r = -3$)

ii. To change $r = 4\sec(\theta)$ from polar form to Cartesian form, rewrite $r = 4\sec(\theta)$ as $r = \dfrac{4}{\cos(\theta)} \to r\cos(\theta) = 4 \to x = 4$.

C. The most common polar equations that you must know how to graph without a calculator are:

---

1. Line

- $\theta = a$ (y–axis if $a = \pm\dfrac{\pi}{2}$, x-axis if $a = 0$ or $a = \pm\pi$)
- Vertical: $r = a\sec(\theta)$
- Horizontal: $r = a\csc(\theta)$

Examples:

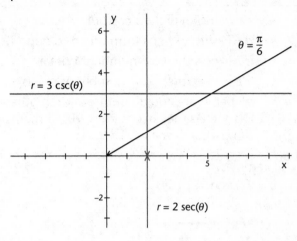

r = 3 csc(θ)

$\theta = \dfrac{\pi}{6}$

r = 2 sec(θ)

2. Circle

- With center at origin: $r = a$, length of radius is a
- Tangent to the y-axis, intersecting the x-axis: $r = a\cos(\theta)$
- Tangent to the x-axis, intersecting the y-axis: $r = a\sin(\theta)$

Examples:

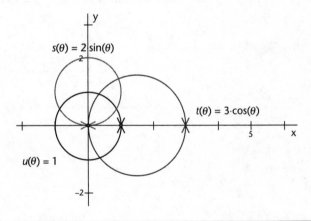

$s(\theta) = 2|\sin(\theta)|$

$t(\theta) = 3 \cdot \cos(\theta)$

$u(\theta) = 1$

3. Rose

- $r = a \sin(b\theta)$ or $r = a \cos(b\theta)$
- a = length of petal from origin to opposite point
- if b is odd then b = number of petals
- if b is even then $2b$ = number of petals
- All petals are equidistant from each other—if there are four petals they occur every 90°, if there are three petals they occur every 120°, etc.

Examples:
$b = 2$, even, 4 petals

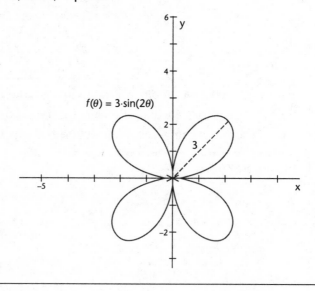

$$f(\theta) = 3 \cdot \sin(2\theta)$$

$b = 3$, odd, 3 petals

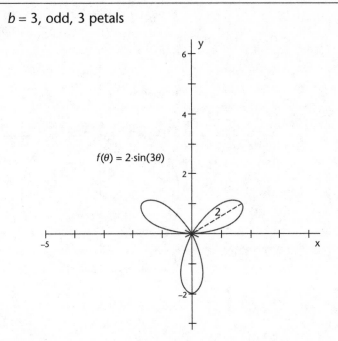

$f(\theta) = 2 \cdot \sin(3\theta)$

4. Limaçon

$r = a \pm b \sin(\theta)$ or $r = a \pm b \cos(\theta)$

- The distance from origin to farthest point from origin is $|a| + |b|$.
- If $|a| > |b|$ the limaçon is dimpled (dimple's distance from the origin is $\big| |a| - |b| \big|$).
- If $|a| < |b|$ the limaçon is looped (length of loop is $\big| |a| - |b| \big|$).
- If limaçon equation contains 'sin' and '+', the graph lies mostly above the x-axis.
- If limaçon equation contains 'sin' and '−', the graph lies mostly below the x-axis.
- If limaçon equation contains 'cos' and '+', the graph lies mostly to the right of the y-axis.
- If limaçon equation contains 'cos' and '−', the graph lies mostly to the left of the y-axis.

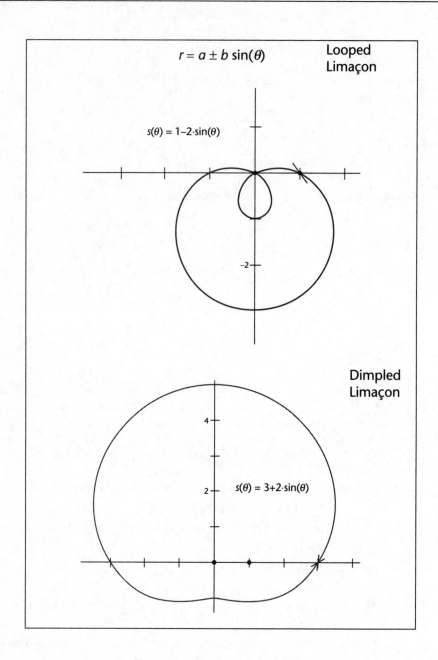

$r = a \pm b \sin(\theta)$

Looped Limaçon

$s(\theta) = 1 - 2 \cdot \sin(\theta)$

Dimpled Limaçon

$s(\theta) = 3 + 2 \cdot \sin(\theta)$

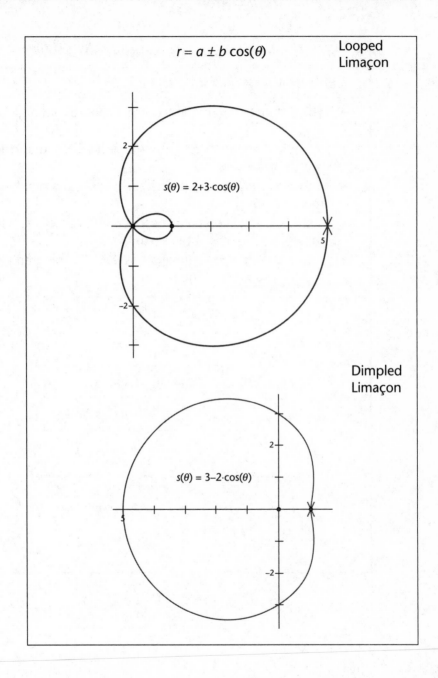

5. Cardioid (type of limaçon in which $a = b$)

$r = a \pm a \sin(\theta)$ or $r = a \pm a \cos(\theta)$

- The distance from origin to farthest point from origin is $2|a|$.
- If cardioid equation contains 'sin' and '+', the graph lies mostly above x-axis.
- If cardioid equation contains 'sin' and '−', the graph lies mostly below the x-axis.
- If cardioid equation contains 'cos' and '+', the graph lies mostly to the right of the y-axis.
- If cardioid equation contains 'cos' and '−', the graph lies mostly to the left of the y-axis.

$r = a \pm a \sin(\theta)$

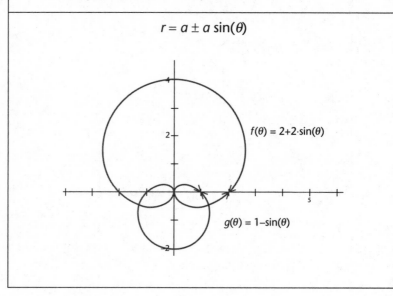

$f(\theta) = 2 + 2 \cdot \sin(\theta)$

$g(\theta) = 1 - \sin(\theta)$

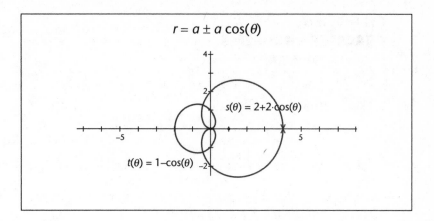

$r = a \pm a \cos(\theta)$

$s(\theta) = 2+2\cdot\cos(\theta)$

$t(\theta) = 1-\cos(\theta)$

Keep in Mind...

➤ Indicate the direction of motion on graphs represented by parametric equations or vector equations.

➤ When graphing parametric equations or vector equations, take into account the restriction on the parameter.

➤ When using $\theta = \tan^{-1}\left(\dfrac{y}{x}\right)$ to calculate the reference angle for a point in Cartesian form, beware of pairing the right r value with the right θ value. For example, to write the Cartesian point $(-1, \sqrt{3})$ in polar form, $r^2 = x^2 + y^2 = 1^2 + (\sqrt{3})^2 = 4 \rightarrow r = \pm 2$ and $\theta = \tan^{-1}\left(\dfrac{\sqrt{3}}{-1}\right) = -\dfrac{\pi}{3}$. So, the point can be represented by $\left(2, \dfrac{2\pi}{3}\right)$ or by $\left(-2, -\dfrac{\pi}{3}\right)$, though there is an infinite number of representations of a point in polar form.

CHAPTER 6
PRACTICE PROBLEMS

(See solutions on page 200)

1. Sketch the graph given by $\begin{cases} x(t) = 2\sin(t) \\ y(t) = 3\cos(t) \end{cases}$ $0 \le t \le \pi$

2. Sketch the graph given by $r(t) = \left(\dfrac{t}{2}\right)i + (e^t)j$ $t > 0$

3. Name and sketch $r = 2 - 3\cos(\theta)$.

4. Find the number of petals and the length of each petal of $r = 4\sin(6\theta)$.

5. Write in Cartesian form: $r = \cos(\theta)$.

6. Write in polar form: $x = 2$.

PART III
DERIVATIVES

Derivatives

I. DERIVATIVES

A. Meaning of Derivative

The <u>derivative</u> of a function is its <u>slope</u>. A linear function has a constant derivative since its slope is the same at every point. The derivative of a function at a point is the slope of its tangent line at that point. Non-linear functions have changing derivatives since their slopes (slope of their tangent line at each point) change from point to point.

1. Local linearity or linearization—when asked to find the linearization of a function at a given x-value or when asked to find an approximation to the value of a function at a given x-value using the tangent line, this means finding the equation of the tangent line at a "nice" x-value in the vicinity of the given x-value, substituting the given x-value into it and solving for y.

   i. For example, approximate $\sqrt{4.02}$ using the equation of a tangent line to $f(x) = \sqrt{x}$. We'll find the equation of the tangent line to $f(x) = \sqrt{x}$ at $x = 4$ (this is the 'nice' x-value mentioned earlier). What makes it nice is that it is close to 4.02 and that $\sqrt{4} = 2$). Since $f'(x) = \dfrac{1}{2\sqrt{x}}$, $f'(4) = \dfrac{1}{2\sqrt{4}} = \dfrac{1}{4}$ so, $m = \dfrac{1}{4}$. Also, $f(4) = 2$. Substituting these values into the equation of the tangent line, $y = mx + b \rightarrow 2 = \dfrac{1}{4}(4) + b \rightarrow b = 1$ so the equation of the tangent line is $y = \dfrac{1}{4}x + 1$. Substituting $x = 4.02$,

$y = 2.005$. A more accurate answer (using the calculator) is $\sqrt{4.02} = 2.004993766$. The linear approximation, 2.005, is very close to this answer. This works so well because the graph and its tangent line are very close at the point of tangency, thus making their y-values very close as well. If you use the tangent line to a function at $x = 4$ to approximate the function's value at $x = 9$, you will get a very poor estimate because at $x = 9$, the tangent line's y-values are no longer close to the function's y-values.

ii. The slope of the secant on (a, b), is often used to approximate the value of the slope at a point inside (a, b). For instance, given the table of values of $f(x)$ below, and given that $f(x)$ is continuous and differentiable, approximate $f'(3)$. You will not be told to use the slope of the secant between two points containing $x = 3$, you'll just have to know to do this. So, $m_{x=3} \approx \dfrac{2.9 - 1.3}{5 - 2} = .5\overline{3}$ or $m_{x=3} \approx \dfrac{2.8 - 1.3}{6 - 2} = .375$. There can be different answers since this is only an approximation.

x	f(x)
2	1.3
3	1.6
5	2.9
6	2.8

B. Notation of Derivative and common terms used to describe it

1. Common notations: $f'(x)$, y', $\dfrac{dy}{dx}$, $\dfrac{d}{dx}(f(x))$

2. Common terms to describe the derivative: instantaneous rate of change, change in y with respect to x, slope.

C. Definition of Derivative

1. Derivative as a function: $f'(x) = \lim\limits_{h \to 0} \dfrac{f(x+h) - f(x)}{h}$

2. Derivative at a point, $x = a$: $f'(a) = \lim\limits_{x \to a} \dfrac{f(x) - f(a)}{x - a}$ (Notice that this is equivalent to $m_{tangent@x=a} = \lim\limits_{x \to a}(m_{secant \ between \ x \ and \ a})$.

This is to say that the slope of the tangent line at $x = a$ is equal to the limit of the slope of the secant line between $x = a$ (and any other x-value as the x-value approaches a.).

D. Existence of Derivative at a point

A function's derivative does not exist at points where the function has a discontinuity, corner, cusp, vertical asymptote or vertical tangent.

1. $y = f(x)$

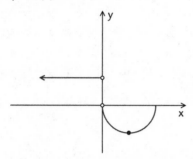

$f'(0)$ does not exist because $f(x)$ is <u>discontinuous</u> at $x = 0$.

2. $y = g(x)$

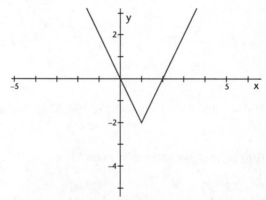

$g'(1)$ does not exist because $g(x)$ has a <u>corner</u> at $x = 1$ and the left and right derivatives are not equal.

3. $y = h(x)$

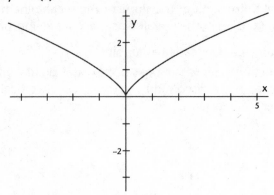

$h'(0)$ does not exist because at $x = 0$ $h(x)$ has a <u>cusp</u> (also a vertical tangent)

4. $y = s(x)$

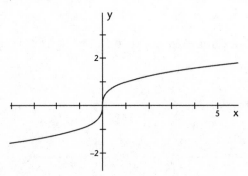

$s'(0)$ does not exist because $s(x)$ has a <u>vertical tangent</u> at $x = 0$

E. Properties of $f(x) = e^x$ and $g(x) = \ln(x)$
   1. $e^{\ln(x)} = x$
   2. $\ln(e^x) = x$
   3. $\ln(xy) = \ln(x) + \ln(y)$
   4. $\ln\left(\dfrac{x}{y}\right) = \ln(x) - \ln(y)$
   5. $\ln(x^y) = y\ln(x)$
   6. $\ln(1) = 0$
   7. $\ln(0)$ does not exist

F. L'Hôspital's Rule—allows you to take limits that have indeterminate forms, such as $\dfrac{0}{0}$ or $\dfrac{\infty}{\infty}$. So, if $\lim\limits_{x \to a} \dfrac{f(x)}{g(x)}$ equals one of these indeterminate forms, then take $\lim\limits_{x \to a} \dfrac{f'(x)}{g'(x)}$. Note that you are <u>not</u> using the quotient rule here, you are simply taking the derivative of the numerator and denominator separately. If the limit still has an indeterminate form then repeat the process as necessary. This also applies to cases in which $x \to \pm\infty$.

1. For example, $\lim\limits_{x \to \infty} \dfrac{x^3}{e^x} = \dfrac{\infty}{\infty}$. Using L'Hôspital's rule,

$$\lim_{x \to \infty} \frac{x^3}{e^x} = \lim_{x \to \infty} \frac{3x^2}{e^x} = \lim_{x \to \infty} \frac{6x}{e^x} = \lim_{x \to \infty} \frac{6}{e^x} = 0.$$

G. Derivative Rules

y	y'	Rule Name	Examples
k	0	Constant Rule	$y = 3 \to y' = 0$
ax^n	anx^{n-1}	Power Rule	$y = 3x^5 - 2x^4 \to y'$ $= 15x^4 - 8x^3$
$f(x) \cdot g(x)$	$f(x)\,g'(x) + g(x)f'(x)$	Product Rule	$y = x^2(x^3 - 3x) \to y'$ $= x^2(3x^2 - 3)$ $+ (x^3 - 3x)(2x)$
$\dfrac{f(x)}{g(x)}$	$\dfrac{g(x)f'(x) - f(x)g'(x)}{[g(x)]^2}$	Quotient Rule	$y = \dfrac{x^2}{2x+1} \to y'$ $= \dfrac{(2x+1)(2x) - x^2(2)}{(2x+1)^2}$ $= \dfrac{2x^2 + 2x}{(2x+1)^2}$

y	y'	Rule Name	Examples
$f(g(x))$	$f'(g(x))g'(x)$	Chain Rule	$y = (x^2 + 3x + 7)^5 \rightarrow y'$ $= 5(x^2 + 3x + 7)^4(2x + 3)$ Here, the "inside function," $g(x) = (x^2 + 3x + 7)$ The "outside function," $f(x) = x^5$.

1. When taking the derivative of a function you might have to use more than one of the above rules.

2. There are some functions whose derivatives occur very often on the exam and it would save you time if memorized. These are the derivatives of $y = \dfrac{1}{x} \rightarrow y' = -\dfrac{1}{x^2}$ and more generally, $y = \dfrac{k}{f(x)} \rightarrow y' = -\dfrac{kf'(x)}{[f(x)]^2}$; and $y = \sqrt{x} \rightarrow$ $y' = \dfrac{1}{2\sqrt{x}}$ and more generally, $y = \sqrt{f(x)} \rightarrow y' = \dfrac{f'(x)}{2\sqrt{f(x)}}$.

Note that the chain rule was used in both general cases.

H. Derivatives of trigonometric functions

y	$\sin(x)$	$\cos(x)$	$\tan(x)$	$\cot(x)$	$\sec(x)$	$\csc(x)$
y'	$\cos(x)$	$-\sin(x)$	$\sec^2(x)$	$-\csc^2(x)$	$\sec(x)\tan(x)$	$-\csc(x)\cot(x)$

1. The derivatives of the cofunctions are negative.

2. In taking the derivative of most trigonometric functions you will need to use the chain rule since most will be compositions—sometimes of more than two functions. Here is an example of the derivative of a function of the form $y = f(g(h))$: $y = \sin(\tan(x^2)) \rightarrow y' = \cos(\tan(x^2))\sec^2(x^2)(2x)$.

I. Derivatives of inverse trigonometric functions

y	$\sin^{-1}(x)$	$\cos^{-1}(x)$	$\tan^{-1}(x)$	$\cot^{-1}(x)$	$\sec^{-1}(x)$	$\csc^{-1}(x)$
y'	$\dfrac{1}{\sqrt{1-x^2}}$	$-\dfrac{1}{\sqrt{1-x^2}}$	$\dfrac{1}{1+x^2}$	$-\dfrac{1}{1+x^2}$	$\dfrac{1}{x\sqrt{x^2-1}}$	$-\dfrac{1}{x\sqrt{x^2-1}}$

1. Note that the derivatives of the cofunctions are the negatives of the derivatives of the functions.
2. In most cases, the chain rule is used. For example,

$$y = \sec^{-1}(\cos(3x)) \rightarrow y' = \frac{-3\sin(3x)}{(\cos(3x))\sqrt{\cos^2(3x) - 1}}.$$

J. Implicit Differentiation—this means finding y' when the equation given is not explicitly defined in terms of y (that is, it is not of the form $y = f(x)$). In this case you must remember to always use the chain rule when taking the derivative of an expression involving y. That is all!

Example 1: Find y' if $x^2 + y^2 = 3$.

Taking derivatives on both sides, $2x + 2yy' = 0 \rightarrow y' = -\frac{x}{y}$.

Example 2: Find y' if $x^2y^2 - 3 \ln y = x + 7$.

Taking derivatives on both sides, $x^2(2yy') + y^2 2x - \frac{3}{y}y' =$

$1 \rightarrow 2x^2yy' - \frac{3}{y}y' =$

$1 - 2xy^2 \xrightarrow{\text{factor out } y'} y' = \frac{1 - 2xy^2}{\left(2x^2y - \dfrac{3}{y}\right)}$

$\xrightarrow{\text{multiply numerator and denominator by } y} y' = \frac{y - 2xy^3}{2x^2y^2 - 3}$. Note that the

product rule must be used here when taking the derivative of x^2y^2.

Example 3: Find y' if $x^2 - xy = x + y$.
Taking derivatives on both sides, $2x - (xy' + y) = 1 + y' \rightarrow$
$2x - xy' - y = 1 + y' \rightarrow 2x - y -$

$1 = y' + xy' \rightarrow y' + xy' = 2x - y - 1 \rightarrow y'(1 + x) = 2x - y - 1 \rightarrow$

$y' = \frac{2x - y - 1}{1 + x}$. Note that you must use the product rule

when taking the derivative of xy – and must distribute the negative sign!

K. The derivative of the inverse of a function: $y = f(x) \rightarrow$

$$\frac{d}{dx}(y^{-1}) = \frac{1}{f'(f^{-1}(x))}$$

1. An example using the formula: $f(x) = \sqrt{x}$. Since $f' = \dfrac{1}{2\sqrt{x}}$

   and $f^{-1}(x) = x^2$ (for $x > 0$), then, $\dfrac{d}{dx}(y^{-1}) = \dfrac{1}{f'(f^{-1}(x))} =$

   $\dfrac{1}{\dfrac{1}{2\sqrt{x^2}}} = \dfrac{1}{\dfrac{1}{2x}} = 2x$. This is only an illustration of this formula.

   Certainly you can find the derivative of the inverse more directly by finding the inverse first and then taking its derivative. In many cases this is difficult or impossible or simply time-consuming. Generally, to find the derivative of the inverse of a function, switch x and y and find y' implicitly. If asked to evaluate the derivative of the inverse of a function at a point, make sure you know which point you are given, one on the function or one on the inverse. Remember that if (a, b) is a point on a function, then (b, a) is a point on its inverse. The converse is also true.

2. An example without using the formula: given $f(x) = x^3 + 2$, evaluate $(f^{-1}(3))'$. Notice that $x = 3$ is an x-value of the inverse. So, rewrite the function, $y = x^3 + 2$, switch x and y, $x = y^3 + 2$. To find y' take the derivative implicitly:

   $1 = 3y^2 y' \rightarrow y' = \dfrac{1}{3y^2}$. The y in this equation is the y of the

   inverse. So, since $x = 3$, $y = 1$ (substitute $x = 3$ into $x = y^3 + 2$

   to find the y-value) and our final answer is $(f^{-1}(3))' = \dfrac{1}{3}$.

L. Derivatives of natural log and exponential functions.

1. $y = \ln(x) \rightarrow y' = \dfrac{1}{x}$, $x > 0$. In general, using the chain rule,

   $y = \ln(f(x)) \rightarrow y' = \dfrac{f'(x)}{f(x)}$, $f(x) > 0$.

2. $y = a^x \rightarrow y' = a^x \ln a$. In general, using the chain rule,
   $y = a^{f(x)} \rightarrow y' = a^{f(x)} f'(x) \ln a$. A special case of this is $y = e^x$.
   It is the only function which is equal to its derivative,
   $(e^x)' = e^x$ other than the trivial $y = 0$.

3. $y = f(x)^{g(x)}$. The formula for this is too long. It rarely appears on the AP exam but, to be safe, know how to find its derivative—by combining logarithmic and implicit differentiation. This is performed when the exponent is a variable. For example:

$$y = (3x)^{x+2} \rightarrow \ln(y) = (x+2)\ln(3x) \rightarrow \frac{1}{y}y' = (x+2)\left(\frac{1}{x}\right)$$

$$+ \ln(3x)(1) \rightarrow y' = y\left[(x+2)\left(\frac{1}{x}\right) + \ln(3x)\right] \rightarrow y'$$

$$= (3x)^{x+2}\left[(x+2)\left(\frac{1}{x}\right) + \ln(3x)\right].$$

M. Derivatives of piecewise functions

1. In order for a piecewise function to be differentiable at a break point, it must be continuous at that point, and the derivatives of the pieces at that point must be equal. Remember that differentiability implies continuity but continuity does not imply differentiability. This is to say that if a function is differentiable at a point, then it is continuous at that point. However, if a function is continuous at a point, it is not necessarily differentiable at that point.

i. $y = \begin{cases} x^2 + 1, & x < 1 \\ 2x, & x > 1 \end{cases}$. This function is not differentiable at

$x = 1$ because it is not continuous there. So,
$$y' = \begin{cases} 2x, & x < 1 \\ 2, & x > 1 \end{cases}.$$

ii. $y = \begin{cases} x^2 + 1, & x \leq 1 \\ 2x, & x > 1 \end{cases}$. This function is continuous at $x = 1$

because $(1)^2 + 1 = 2(1)$ and the derivatives of the pieces are equal at $x = 1$ ($2x = 2$ at $x = 1$). Therefore, this function is differentiable at $x = 1$ and $y' = \begin{cases} 2x, & x \leq 1 \\ 2, & x > 1 \end{cases}.$

iii. $y = \begin{cases} x^2, & x \le 1 \\ 2x, & x > 1 \end{cases}$. This function is not differentiable at $x = 1$

because, even though the derivatives of the pieces at $x = 1$ are equal, this function is not continuous at $x = 1$.

So it is not differentiable there. $y' = \begin{cases} 2x, & x < 1 \\ 2, & x > 1 \end{cases}$

Notice that $x = 1$ was excluded from the domain of the derivative since the derivative does not exist there.

iv. $y = \begin{cases} x^2 + 1, & x \le 1 \\ 2, & x > 1 \end{cases}$. This function is continuous at $x = 1$

but not differentiable there since $2x \ne 0$ when $x = 1$. That is, the derivatives of the pieces aren't equal there.

N.* Derivatives of parametric and polar equations

1. The first derivative—given $\begin{cases} x(t) = f(t) \\ y(t) = g(t) \end{cases} \rightarrow \dfrac{dy}{dx} = \dfrac{dy/dt}{dx/dt}$. For

   example, $\begin{cases} x(t) = 2t^2 + 3t \\ y(t) = e^{2t} \end{cases} \rightarrow \dfrac{dy}{dx} = \dfrac{2e^{2t}}{4t + 3}$. Note that the

   derivative of a set of parametric equations is a function of t.

Test Tip

Remember that this derivative represents the slope of the graph at a given x value. But, when asked to find the slope at a given x value, substitute it back in to the original to find the t value corresponding to it. Then use that t value to substitute into the derivative and find the slope! The common mistake is to use the x value instead of the t value.

2. The second derivative, $\dfrac{d^2y}{dx^2} = \dfrac{dy'/dt}{dx/dt}$. In words, the

   numerator of this formula represents the second derivative of $y(t)$ and the denominator represents the first derivative of $x(t)$. Since

   $\dfrac{dy}{dx} = \dfrac{2e^{2t}}{4t+3}, \dfrac{dy'}{dt} = \dfrac{d}{dt}\left(\dfrac{2e^{2t}}{4t+3}\right) = \dfrac{(4t+3)(4e^{2t}) - (4)(2e^{2t})}{(4t+3)^2} =$

   $\dfrac{16te^{2t} + 4e^{2t}}{(4t+3)^2} = \dfrac{4e^{2t}(4t+1)}{(4t+3)^2}$. Thus, $\dfrac{d^2y}{dx^2} = \dfrac{4e^{2t}(4t+1)}{(4t+3)^2} \div (4t+3) = \dfrac{4e^{2t}(4t+1)}{(4t+3)^3}$.

*This is a BC Calculus topic.

The most common mistake in finding the second derivative is to use the quotient rule to find the derivative of the first derivative. It does not work that way!

O.* Derivatives of polar equations

1. Rewrite the polar equations in parametric form and use the parametric formulas.

   i. First derivative—rewrite $r = f(\theta)$ in parametric form:
   $$\begin{cases} x(\theta) = r\cos(\theta) \\ y(\theta) = r\sin(\theta) \end{cases} \rightarrow \begin{cases} x(\theta) = f(\theta)\cos(\theta) \\ y(\theta) = f(\theta)\sin(\theta) \end{cases}$$

   $$\frac{dy}{dx} = \frac{dy/d\theta}{dx/d\theta} = \frac{f(\theta)\cos(\theta) + \sin(\theta)f'(\theta)}{-f(\theta)\sin(\theta) + \cos(\theta)f'(\theta)}.$$ For example,

   find the derivative (slope) of $f(\theta) = 2\sin(3\theta)$ at $\theta = \dfrac{\pi}{6}$

   $$\frac{dy}{dx} = \frac{dy/d\theta}{dx/d\theta} = \frac{f(\theta)\cos(\theta) + \sin(\theta)f'(\theta)}{-f(\theta)\sin(\theta) + \cos(\theta)f'(\theta)}$$

   $$= \frac{2\sin(3\theta)\cos\theta + \sin(\theta)6\cos(3\theta)}{-2\sin(3\theta)\sin(\theta) + \cos(\theta)6\cos(3\theta)} \rightarrow \left.\frac{dy}{dx}\right|_{\theta = \frac{\pi}{6}} = -\sqrt{3}.$$

   Graphically, this represents the slope of the tangent line

   to the graph of $f(\theta) = 2\sin(3\theta)$ at the point $\left(2, \dfrac{\pi}{6}\right)$.

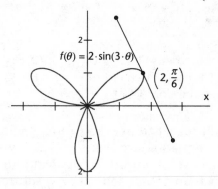

*This is a BC Calculus topic.

ii. Second derivative of a polar equation is found using the parametric formula for the second derivative, though it does not appear on the AP Calculus exams.

Keep in Mind...

➤ Whens using the quotient rule, do not switch the order of the terms in the numerator since subtraction is not commutative.

➤ When differentiating a function of the form $y = \dfrac{f(x)}{k}$ where k is a constant, do not use the quotient rule. The derivative is simply $y' = \dfrac{f'(x)}{k}$ since $\dfrac{1}{k}$ is a constant that can be factored out.

➤ When asked to find the derivative of parametric equations at a certain point, pay attention to whether you are given an x-value or a t-value and solve the problem accordingly.

➤ The second derivative of parametric equations is tricky; make sure you don't fall for it! Practice it until you get it right.

➤ Do not confuse $\ln(1) = 0$ with $\ln(0) = 1$. The former is true since $x = 1$ is the x-intercept of $y = \ln(x)$. The latter is false since $x = 0$ is not in the domain of $y = \ln(x)$.

➤ Don't forget to use the product rule in implicit differentiation problems in which you must take the derivative of a product involving both x and y. Also, if such an expression is being subtracted, make sure to distribute the negative sign.

CHAPTER 7
PRACTICE PROBLEMS

(See solutions on page 202)

1. Find $\dfrac{dy}{dx}$ if $y = \dfrac{3x+4}{1-5x}$

2. Find $\dfrac{d}{dx}(\ln(e^{\sqrt{5x+3}}))$

3. Evaluate y' at $x = -1$ if $3x - x^2y = 5y$

4. Find the derivative of the inverse of $y = x^2 - 4x$ at $x = 2$.

Curve Sketching

I. SKETCHING $f(x)$ GIVEN ITS EQUATION

A. Derivatives and intervals of increase and decrease

1. If $f'(x) > 0$ on (a, b) then $f(x)$ is increasing on (a, b).
   Ex: $f(x) = x^2$ on $(0, \infty)$

2. If $f'(x) < 0$ on (a, b), then $f(x)$ is decreasing on (a, b),
   Ex: $f(x) = x^2$ on $(-\infty, 0)$

3. If $f'(x) = 0$ at $x = a$ then $x = a$ is a candidate for the x-value
   of a max/min point. Ex: $f(x) = x^2$ at $x = 0$ there's a minimum
   point because $f'(0) = 0$ and f' changes sign from negative to
   positive here. Ex: At $x = 0$, $f(x) = x^3$ does not have a max or
   a min point because, even though $f'(0) = 0$, $f'(x)$ does not
   change sign here.

4. If $f'(x)$ dne at $x = a$ and $f(a)$ exists, then $x = a$ is a candidate
   for the x-value of a max/min point.

   i. At $x = 0$, $f(x) = x^{\frac{2}{3}}$ has an absolute minimum point because
      $f'(0)$ dne, $f(0)$ exists, and $f'(x)$ changes sign from negative
      to positive.

   ii. At $x = 0$, $f(x) = x^{\frac{1}{3}}$ does not have a max or min point
       because, though $f'(0)$ does not exist and $f(0)$ exists, $f'(x)$
       does not change sign here.

5. Critical points of $f(x)$ are points in its domain at which
   $f'(x) = 0$ or $f'(x)$ does not exist.

B. Derivatives and concavity

1. If $f''(x) > 0$ on (a, b), then $f(x)$ is concave up on (a, b).
   Ex: $f(x) = x^2$ on $(-\infty, \infty)$

2. If $f''(x) < 0$ on (a, b), then $f(x)$ is concave down on (a, b).
   Ex: $f(x) = -x^2$ on $(-\infty, \infty)$

3. If $f''(x) = 0$ then $x = a$ is a candidate for the x-value of an *inflection point* (a point at which the concavity of $f(x)$ changes).

   i. $f(x) = x^3$ has an inflection point at $x = 0$ because $f''(0) = 0$ and $f''(x)$ changes sign from negative to positive. An inflection point occurs at $x = a$ if and only if $f(a)$ exists, $f''(a) = 0$ or does not exist, **and** $f''(x)$ changes signs!

C. Graphing a function requires finding its intercepts, relative and absolute extrema, and asymptotes.

   1. Intercepts—an x-intercept is a point at which a function intersects the x-axis, and hence, $y = 0$ here. A y-intercept is a point at which a function intersects the y-axis, and hence, $x = 0$ here. Not all functions have intercepts, for example, $y = \dfrac{1}{x}$. Some functions have an infinite number of x-intercepts, for example, $y = \sin(x)$. A function can have *at most* one y-intercept. If a graph has more than one y-intercept, it violates the definition of function because it would have more than one different y-value for $x = 0$. No calculus is necessary to find intercepts.

   2. Relative maximum/minimum points—a point on a function is a relative (or local) maximum point if and only if it is the highest point in its neighborhood. Think of it as the top of a mountain but not necessarily the highest mountain. A point on a function is a relative (or local) minimum point if and only if it is the lowest point in its neighborhood. Think of it as the bottom of a valley but not necessarily the lowest valley. Relative (or local) extrema (that is, maximum or minimum points) occur at interior points of a function, not at end points. Not all functions have relative extrema, for example, $y = \dfrac{1}{x}$. The relative extrema occur at points where the first derivative is either zero or nonexistent *and* the function is defined. In particular, if the minimum of a function occurs at an interior point of the function, $x = a$, the derivative is negative to the left of a and positive to the right of a—that is, the function must change from decreasing to increasing at

$x = a$. If the maximum of a function occurs at an interior point of the function, $x = b$, the derivative is positive to the left of b and negative to the right of b—that is, the function must change from increasing to decreasing at $x = b$.

3. Absolute maximum/minimum points—a point on a function is an absolute maximum point if and only if it is the highest point. Think of it as the top of the highest mountain. A point on a function is an absolute minimum point if and only if it is the lowest point. Think of it as the bottom of the lowest valley. Absolute extrema can occur at interior points or at endpoints of a function. The absolute extrema occur at points where the first derivative is either zero or nonexistent *and* the function is defined—or at endpoints of the function.

4. Critical points—these are points in the domain of a function at which the derivative is either equal to zero or does not exist. These are generally found when looking for max/min points.

5. Asymptotes—refer to chapter 4.

6. Graphs to illustrate curve sketching

   i.

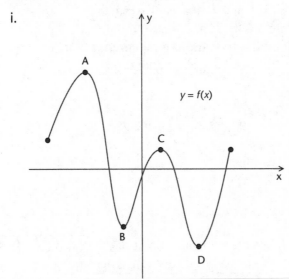

Point A is the absolute maximum point.
Point C is a relative maximum point.
Point B is a relative minimum point.
Point D is the absolute minimum point.

ii.

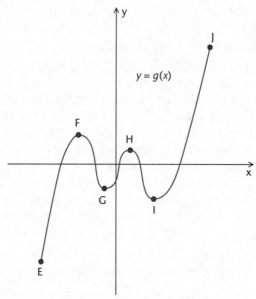

$y = g(x)$

Point J is the absolute maximum point,
but not a relative maximum.
Points F and H are relative maxima.
Points G and I are relative minima.
Point E is the absolute minimum point,
but not a relative minimum.

iii.

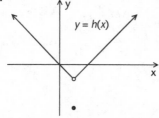

$y = h(x)$

The absolute minimum point of $y = h(x)$ is (1, −3). At $x = 1$,
$h(x)$ is defined but its derivative does not exist.

Test Tip

*A point on a function consists of both the x and y values of
the point, so when asked to find a minimum/maximum point,
find both the x and y values of it. When asked to find the
minimum/maximum value of a function, find only the y-value.*

iv. Find the absolute minimum point of $f(x) = xe^x$.

Step 1: Find $f'(x)$. $f'(x) = xe^x + e^x$

Step 2: Set $f'(x) = 0$ and also check for points where $f'(x)$ does not exist. $f'(x) = xe^x + e^x = 0 \rightarrow e^x(x+1) = 0 \rightarrow x = -1$ (e^x is positive for all values of x). This function has no points of nondifferentiability.

Step 3: Check to see if there is a max or min at $x = -1$ by making a sign analysis chart for $f'(x)$. Make sure to include the x-values found in step 2, then a value from the right and a value from the left of those x-values.

$f'(x)$	negative	**0**	positive
x	-2	**-1**	0

Since $f(-1)$ exists, $f'(-1) = 0$ and $f'(x)$ changes from negative to positive at $x = -1$, we conclude that the absolute minimum point of $f(x)$ occurs at $x = -1$.

Step 4: Since asked to find the absolute minimum *point* of $f(x)$, find the value of $f(-1) = (-1)e^{-1} = -\dfrac{1}{e}$. Final answer: $\left(-1, -\dfrac{1}{e}\right)$. The graph of $f(x) = xe^x$:

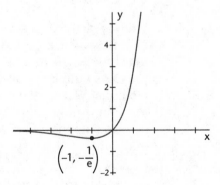

v. Find the absolute maximum and minimum values of
   $g(x) = -x^4 - 2x^3$ on $[-2, 2]$

   Step 1: Find $g'(x)$. $g'(x) = -4x^3 - 6x^2$

   Step 2: Set $g'(x) = 0$ and look for points of
   nondifferentiability. $g'(x) = -4x^3 - 6x^2 = 0 \rightarrow$

   $-2x^2(2x + 3) = 0 \rightarrow x = 0$ or $x = -\dfrac{3}{2}$. $g(x)$ is a

   polynomial so it has no points of nondifferentiability.

   Step 3: Make a sign analysis chart for $g'(x)$ making sure to
   include the endpoints since they are candidates
   for absolute extrema.

$g'(x)$	positive	**0**	negative	**0**	negative
x	-2	$-\dfrac{3}{2}$	-1	0	2

   Since $g'(x)$ changes from positive to negative

   at $x = -\dfrac{3}{2}$, and $g\left(-\dfrac{3}{2}\right)$ exists, $g(x)$ must have a

   maximum here.
   Evaluate the original function at the critical points,

   $\left(x = -\dfrac{3}{2}, x = 0\right)$ and at the endpoints to find

   absolute extrema: $g\left(-\dfrac{3}{2}\right) = -\left(-\dfrac{3}{2}\right)^4 - 2\left(-\dfrac{3}{2}\right)^3 =$

   1.6875; $g(0) = 0$; $g(-2) = 0$; $g(2) = -32$. The
   highest y-value is $y = 1.6875$ so this is the absolute
   maximum value of $g(x)$. The lowest y-value is
   $y = -32$ so this is the absolute minimum of $g(x)$.
   Note that you were *not* asked to find the absolute
   maximum and minimum *points* of $g(x)$ but only
   the absolute maximum and minimum values
   of $g(x)$.

vi. Find the critical points of $f(x) = \sqrt{x}$ and characterize them
   as absolute maximum or absolute minimum points.

   Step 1: Find y'. $y' = \dfrac{1}{2\sqrt{x}}$

   Step 2: In this case, note that $f'(x)$ does not equal zero for
   any x-value and $f'(x)$ does not exist at $x = 0$. Since

$f(0)$ is defined, there is a critical point at $x = 0$ and this critical point is a candidate for absolute max/min.

Step 3: Make a sign analysis chart for $f'(x)$.

$f'(x)$	**dne**	positive
x	**0**	2

Note that we cannot include negative values in the sign analysis chart since these are not included in the domain of $f(x)$.

From this chart we conclude that the function, $y = \sqrt{x}$, increases without bound since its derivative is positive on $(0, \infty)$.

Step 4: Since y is increasing on $x > 0$, the lowest y value must occur at $x = 0$ and $y\big|_{x=0} = 0$. According to the sign analysis chart for $f'(x)$, $f(x)$ has only one absolute minimum point at $(0, 0)$, its left endpoint.

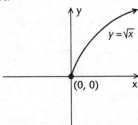

vii. Given $f(x) = \dfrac{x^2 - 1}{x^2 - 4}$ find its intercepts, asymptotes, intervals of increase and decrease, relative max/min points, absolute max/min points, concavity intervals, points of inflection and graph it.

<u>Intercepts:</u>

x-intercept, set $y = 0$: $\dfrac{x^2 - 1}{x^2 - 4} = 0 \to x^2 - 1 = 0 \to$

$x = \pm 1$ (Note that the denominator does not equal zero at $x = \pm 1$.)

y-intercept: set $x = 0$: $\dfrac{0^2 - 1}{0^2 - 4} = \dfrac{1}{4} \to y = \dfrac{1}{4}$

Vertical asymptotes: $x^2 - 4 = 0 \rightarrow x = \pm 2$. Note that only the denominator equals zero at $x = \pm 2$, not the numerator.

Horizontal asymptotes: $\lim\limits_{x \to \infty} \dfrac{x^2 - 1}{x^2 - 4} = 1$ and $\lim\limits_{x \to -\infty} \dfrac{x^2 - 1}{x^2 - 4} = 1 \rightarrow$

Horizontal asymptote: $y = 1$.

Note: If the limits were equal to different numbers, then the function would have two different horizontal asymptotes. If one limit were equal to a number, $y = b$, and the other to plus or minus infinity, $y = b$ would be the only horizontal asymptote. If both limits were equal to plus or minus infinity, then the function would not have any horizontal asymptotes.

<u>Intervals of increase/decrease and max/min points</u>

$f'(x) = -\dfrac{6x}{(x^2 - 4)^2} = 0$ at $x = 0$. $f'(x) = -\dfrac{6x}{(x^2 - 4)^2}$ does not

exist at $x = \pm 2$

$f'(x)$	positive	dne	positive	0	negative	dne	negative
x	-3	-2	-1	0	1	2	1

According to the sign analysis chart for $f'(x)$, $f(x)$ is increasing on $(-\infty, -2) \cup (-2, 0)$ and decreasing on $(0, 2) \cup (2, \infty)$. Relative maximum point occurs at $\left(0, \dfrac{1}{4}\right)$.

<u>Intervals of concavity and inflection points</u>

$f''(x) = \dfrac{18x^2 + 24}{\left(x^2 - 4\right)^3} \neq 0$ since the numerator is positive for

all x values. $f''(x) = \dfrac{18x^2 + 24}{\left(x^2 - 4\right)^3}$ does not exist at $x = \pm 2$

but since $f(\pm 2)$ does not exist, there are no inflection points for $f(x)$.

$f''(x)$	positive	**dne**	negative	**dne**	positive
x	-3	**-2**	0	**2**	3

According to the sign analysis chart for $f''(x)$, $f(x)$ is concave up on $(-\infty, -2) \cup (2, \infty)$ and concave down on $(-2, 2)$.

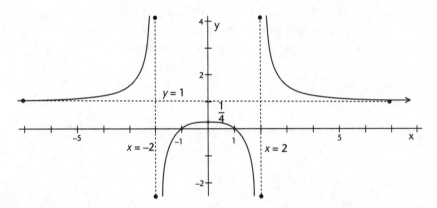

II. SKETCHING $f(x)$ GIVEN THE GRAPH OF $f'(x)$

A. When given the graph of $f'(x)$ create a sign analysis chart for $f'(x)$ and then draw $f(x)$ based on it and any other information given. Analyzing the slopes of the graph of $f'(x)$ also helps you to find the concavity intervals of $f(x)$.

   1. Given the graph of $f'(x)$ below, and $f(-4) = f(5) = -1$ and $f(0) = 3$, sketch the graph of $f(x)$.

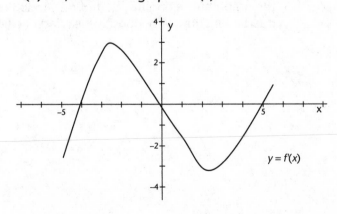

$y = f'(x)$

x	$x < -4$	-4	$-4 < x < -2.5$	-2.5	$-2.5 < x < 0$	0	$0 < x < 2.5$	2.5	$2.5 < x < 5$	5	$x > 5$
$f'(x)$	$-$	zero	$+$	max	$+$	zero	$-$	min	$-$	zero	$+$
$f''(x)$	$+$	$+$	$+$	zero	$-$	$-$	$-$	zero	$+$	$+$	$+$
$f(x)$	decreasing concave up	minimum concave up	increasing concave up	inflection point	increasing concave down	maximum concave down	decreasing concave down	inflection point	decreasing concave up	minimum concave up	increasing concave up

According to the sign analysis charts for $f'(x)$ and $f''(x)$, the graph of $f(x)$ looks like the following:

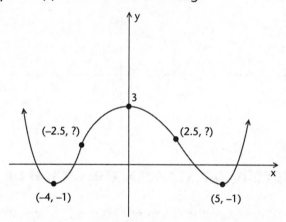

Note 1: The y-value of the inflection point is not known and it is not necessary for sketching the original function.

Note 2: The graph of the derivative looks like a cubic function with a positive leading coefficient. Therefore, expect the graph of the original, that of $f(x)$, to look like a quartic function with a positive leading coefficient.

III. MOTION

A. Rectilinear Motion—motion in a straight line

1. Displacement and distance

   i. Displacement (vector) is the distance between a moving object's end point and starting point. For instance, if an object moves from the origin two units to the right and back to the origin, its displacement is zero. If it moves from the origin two units to the right and then three units to the left, its total displacement is $-1- 0 = -1$ unit. Symbolically, if $S(t)$ represents the path of an object then the displacement from $t = a$ to $t = b$ is given by $S(b) - S(a)$.

   ii. Distance (scalar) is the length of the path traveled by an object. If an object moves from the origin two units to the right and back to the origin, the distance it traveled is 4 units. If it moves from the origin two units to the right and then three units to the left, the distance traveled is 5 units.

2. Speed, velocity, and acceleration

   i. Speed (scalar) measures an object's change in *distance* traveled per unit of time.

   ii. Velocity (vector) measures an object's *displacement* per unit of time. Symbolically, speed = |velocity|. If $s(t)$ represents the displacement of an object, then $s'(t)$ represents the object's velocity.

   iii. Acceleration (vector) measures an object's change in velocity per unit of time. Symbolically, $a(t) = v'(t) = s''(t)$. If the acceleration and velocity of an object have the same sign, then the object is speeding up. If the acceleration and velocity have opposite signs, the object is slowing down. Think of acceleration and velocity as two forces acting on the object—if they act in the same direction, they increase the object's speed; if they act in different directions, they slow the object down.

   Example. The graph of $s(t)$, below, represents an object's displacement. From this graph we can deduce that the object is speeding up in two different ways. Notice that the slope of the tangent line to s(t); that is, the velocity, increases as time increases so the object is speeding up. Or, we can analyze the signs of the velocity and acceleration. In this case, $v(t) > 0$ (since $s(t)$ is increasing) and $a(t) > 0$ (since $s(t)$ is concave up), hence, the object is speeding up.

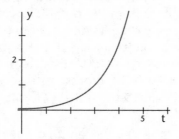

B.* Motion along a parametric/polar curve

1. For parametric curves $x(t)$ and $y(t)$, the velocity vector is $|\langle x'(t), y'(t)\rangle|$, the acceleration vector is $|\langle x''(t), y''(t)\rangle|$ and the speed of the object is the magnitude of the velocity vector,
    $$\text{speed} = |\langle x'(t), y'(t)\rangle| = \sqrt{(x'(t))^2 + (y'(t))^2}.$$

2. For a polar curve $r = f(\theta)$, you must rewrite the original function in parametric form and use the formulas above.

Keep in Mind...

➤ Solutions of $f'(x) = 0$ or $f'(x)$ *dne* give you x-values of possible critical points of $f(x)$, whereas solutions of $f''(x) = 0$ or $f''(x)dne$ give you x-values of possible inflection points.

➤ When asked to find a point, find both x and y values; when asked to find the value of the function, find only the y-value.

*This is a BC Calculus topic.

➤ *Maxima* is the plural of maximum. *Minima* is the plural of minimum.

➤ Remember that critical points and inflection points must be in the domain of the function.

➤ When finding inflection points, don't forget to look for the sign change!

➤ A function can decrease at an increasing rate (a function which is decreasing and concave up). A function can increase at a decreasing rate (a function which is increasing and concave down). A function can increase at an increasing rate (a function which is increasing and concave up). A function can decrease at a decreasing rate (a function which is decreasing and concave down).

CHAPTER 8
PRACTICE PROBLEMS
(See solutions on page 203)

1. Find the critical points, inflection points, the absolute minimum value of y, and relative maximum points of $y = x^4 - 3x^2 + 2$.

2. Sketch the graph of $f(x)$ if the graph of $f'(x)$ is given below:

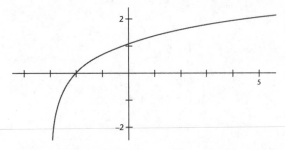

3. The path of a particle is described by the equation $\langle x(t), y(t) \rangle = \langle e^{-2t}, 3t^2 \rangle$. Find the velocity and speed of the particle at $t = 0$.

Optimization and Related Rates

I. OPTIMIZATION

A. Optimizing a quantity means to find its maximum or minimum point. For instance, one could find the maximum profit or the minimum loss in a business situation. These are word problems so they must be read carefully. The steps to solve an optimization problem are:

(a) Create a legend that includes the given information and the variable that you are looking for.

(b) Write down the function that needs to be optimized.

(c) Take the derivative of the function in part b, set it equal to zero, and solve. Justify, using a sign analysis chart, that you found a maximum or a minimum, as the case may be.

(d) Double-check that you found the answer to the question being asked.

 Test Tip

Make sure to include correct units! If the optimization problem is in the free-response section, write your answer in a complete sentence.

Example: Find the radius of the largest cylinder that can be inscribed in a cone of radius 3 in. and height 5 in.

(a) Create legend: $r_{cone} = 3$ in., $h_{cone} = 5$ in., $r_{cylinder} = ?$ such that $V_{cylinder}$ is maximum? (Optimize $V_{cylinder}$)

(b) The function to be optimized: $V_{cylinder} = \pi r^2_{cylinder} \, h_{cylinder}$. Rewrite $h_{cylinder}$ in terms of $r_{cylinder}$ so the function contains only one variable and can be more easily differentiated. In this case, use the fact that the ratios of corresponding sides of similar triangles are equal. That is,

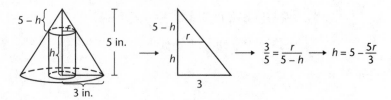

So, $V_{cylinder} = \pi r_{cylinder}^2 \left(5 - \dfrac{5r_{cylinder}}{3} \right) = 5\pi r_{cylinder}^2 - \dfrac{5}{3}\pi r_{cylinder}^3$

(c) Take the derivative of the function in part b) and set it equal to zero: $\dfrac{dV_{cylinder}}{dr} = 10\pi r_{cylinder} - 5\pi r_{cylinder}^2 = 0 -$

$r_{cylinder} = 2\,\text{in.}$ (disregard $r_{cylinder} = 0$). Sign analysis chart for $V'_{cylinder}$ shows that when $r_{cylinder} = 2\,\text{in.}$, $V_{cylinder}$ is a maximum (largest).

$r_{cylinder}$	1	2	3
$\dfrac{dV_{cylinder}}{dr}$	positive	0	negative

(d) The radius of the largest cylinder that can be inscribed in a cone of radius 3 in. and height 5 in. is 2 in.

II. RELATED RATES

A. These are also word problems, very similar to the optimization ones in the sense that you need to take the derivative of a function. The difference being that, in this case, you take the derivative implicitly, with respect to time. If the question states that a snowball is melting at 4 in.³/sec., this means that $\dfrac{dV}{dt} = -4\ \text{in.}^3/\text{min.}$ (look at the units to figure out what variable is being discussed; note that the variable is negative since the snowball's volume is decreasing).

Test Tip

Common formulas you must memorize for such problems are: the Pythagorean Theorem, proportions in right triangles, area/ perimeter of basic geometric figures, and volume of a sphere, cylinder, and cone.

The steps required to solve a related rates problem are:

(a) Create a legend which includes the given information and the variable that you are looking for—it is generally a rate ("how fast is the radius changing with respect to time" means that you are looking for the value of $\frac{dr}{dt}$).

(b) Write the equation that relates the variables given (If the problem involves the radius, height, and volume of a cylinder, the equation you would use would be that for the volume of a cylinder.)

(c) Take the derivative of the function in part b, substitute in the given information and solve for the missing variable.

(d) Double-check that you found the answer to the question being asked. Make sure to include correct units (units squared for area, units cubed for volume). If the related rate problem shows up in the free-response question, write a complete sentence as your answer.

Example: Coffee is poured into a conical cup at a constant rate of 1 in.³/sec. Given that the cup's radius measures 3 in. and its height is 9 in., find how fast the water level of the coffee in the cup changes when the radius is 2 in.

(a) Create legend: $\frac{dV_{coffee}}{dt} = 1\,\text{in.}^3/\text{sec.}$ (this is positive because the coffee volume is increasing), $r_{cup} = 3$ in., $h_{cup} = 9$ in., $\frac{dh_{coffee}}{dt} = ?$ (Note that there are two cones in this problem: one is the cup and the other is the shape of the coffee in the cup. The coffee in the cup changes dimensions but the cup's dimensions remain constant, so it's important to differentiate between the dimensions of the cup and those of the coffee.)

3 in.

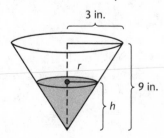

9 in.

r

h

(b) $V_{coffee} = \frac{1}{3}\pi r_{coffee}^2 h_{coffee}$. The volume in this case is a function of two variables. To make it easier to calculate, we must rewrite it so that it is a function of only one variable, h_{coffee}, since we are looking for $\frac{dh_{coffee}}{dt}$. Using the fact that in similar triangles the ratio of corresponding sides forms a proportion, $\frac{r_{coffee}}{h_{coffee}} = \frac{3}{9} \rightarrow r_{coffee} = \frac{1}{3}h_{coffee} \rightarrow$

$$V_{coffee} = \frac{1}{3}\pi\left(\frac{1}{3}h_{coffee}\right)^2 h_{coffee} \rightarrow V_{coffee} = \frac{1}{27}\pi h_{coffee}^3$$

(c) Take the derivative of the function in part (b) and substitute the given information in order to solve for the unknown quantity.

$$\frac{dV_{coffee}}{dt} = \frac{1}{9}\pi h_{coffee}^2 \frac{dh_{coffee}}{dt} \rightarrow 1 = \frac{1}{9}\pi(6)^2 \frac{dh_{coffee}}{dt} \rightarrow$$

$\frac{dh_{coffee}}{dt} = \frac{1}{4\pi}$ in./sec. (since $r_{coffee} = \frac{1}{3}h_{coffee}$, when

$r_{coffee} = 2$ in., $h_{coffee} = 6$ in.)

(d) When the radius of the coffee in the cup is 2 in., the coffee level increases at a rate of $\frac{1}{4\pi}$ in./sec.

Keep in Mind...

➤ Don't forget appropriate units!

➤ If a quantity is increasing, its derivative is positive. If a quantity is decreasing, its derivative is negative. If a quantity does not change, it is constant.

➤ Make sure you know the formulas for the volume of a cone, sphere, cube, and cylinder.

➤ For optimization problems, set the derivative of the function equal to zero. For related rates, substitute the given information into the derivative of the function.

CHAPTER 9
PRACTICE PROBLEMS
(See solutions on page 205)

1. A 13-foot ladder leaning against a wall starts to slip in such a way that the foot of the ladder slips away from the wall at 2 in/sec. How fast is the top of the ladder slipping down the wall when the foot of the ladder is 12 inches from the wall?

2. Find the radius of the largest cylinder that can be inscribed in a sphere of radius 5 inches.

3. A company has x boxes of produce available. The supply equation is given by $px - 10p + 20 = 3x$ where p is the price per box of produce and x is the number of boxes. If x is decreasing at 3 boxes per day, at what rate is the price changing when x is equal to 50 boxes?

The Intermediate-Value Theorem, The Mean Value Theorem, and Rolle's Theorem

I. THE INTERMEDIATE-VALUE THEOREM

The Intermediate-Value Theorem states that a continuous function will take on all values between $f(a)$ and $f(b)$. The graph of a continuous function shown below illustrates the Intermediate-Value Theorem.

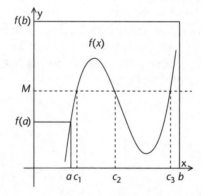

As the graph shows, if we pick any value, M, that is between the value of $f(a)$ and the value of $f(b)$ and draw a line straight out from this point, the line will intersect the graph in at least one point. Thus, somewhere between $x = a$ and $x = b$ the function will take on the value of M. Also, as the figure shows, the function may take on the value at more than one place. In this case, there are three values of c.

Note that the Intermediate-Value Theorem only states that the function will take on the value of M somewhere between a and b. It does not state what that value will be—just that it exists. The Intermediate-Value Theorem is normally used to show that there is a root of a function on a given interval.

Example 1: Show that $f(x) = 2x^3 - 8x^2 + 5x - 2$ has at least one root between $x = 1$ and $x = 4$.

Since a polynomial is continuous and $f(-1) = -17$ and $f(4) = 18$, by the Intermediate-Value Theorem there must be at least one value c on $(-1, 4)$ where $f(c) = 0$.

II. THE MEAN VALUE THEOREM

A. The Mean Value Theorem (MVT) states that if a function is continuous on $[a, b]$ and differentiable on (a, b) then there exists at least one x value, $x = c$, where $a < c < b$, such that $f'(c) = \dfrac{f(b) - f(a)}{b - a}$. In English, this says that if a function is smooth (no breaks, corners or cusps) on an interval, then there must be at least one point in that interval at which the slope of the tangent line equals the slope of the secant line between the endpoints. Equivalently, there must be at least one point in the interval at which the tangent line is parallel to the secant between the end points.

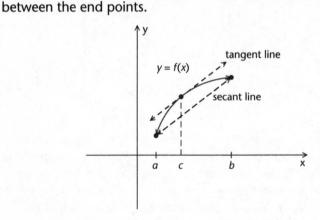

Example 1: Find the value of c guaranteed by the Mean Value Theorem for $f(x) = x^2$ on $[0, 3]$.

Since $f'(x) = 2x$, $f'(c) = 2c$. So, $f'(c) = \dfrac{f(b) - f(a)}{b - a} \rightarrow$

$2c = \dfrac{f(3) - f(0)}{3 - 0} = \dfrac{9}{3} = 3 \rightarrow 2c = 3 \rightarrow c = \dfrac{3}{2}$.

Example 2: Find the value of c guaranteed by the Mean Value Theorem for $f(x) = x^3 - 4x^2 - x + 4$ on $[-1, 2]$.

Since $f'(x) = 3x^2 - 8x - 1$, $f'(c) = 3c^2 - 8c - 1$. So,

$f'(c) = \dfrac{f(b) - f(a)}{b - a} \rightarrow 3c^2 - 8c - 1 = \dfrac{f(2) - f(-1)}{2 - (-1)} = \dfrac{-6}{3} = -2 \rightarrow$

$3c^2 - 8c + 1 = 0 \rightarrow$

$c \approx .131$, or $c \approx 2.535$. Final answer: $c \approx .131$ (reject $c \approx 2.535$ since it is not in the interval given. Also, round off—always at the end of a problem—to at least three decimal places.)

III. ROLLE'S THEOREM

A. Rolle's Theorem states that if a function is continuous on $[a, b]$ and differentiable on (a, b) and $f(a) = f(b)$, then there exists an x value, $x = c$, where $a < c < b$, such that $f'(c) = 0$. This is a simpler case of the Mean Value Theorem in which $f(a) = f(b)$. Clearly, if this is the case, the numerator of the fraction in the MVT becomes zero, thus $f'(c) = 0$.

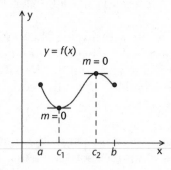

Example 1: Find the value of c guaranteed by Rolle's Theorem for $f(x) = x^3 - 4x^2 - x + 4$ on $[-1, 4]$.

Since $f'(x) = 3x^2 - 8x - 1$, $f'(c) = 3c^2 - 8c - 1$. So, $f'(c) = 0 \rightarrow 3c^2 - 8c - 1 = 0 \rightarrow c \approx -.120$ or $c \approx 2.786$.

 IV. **WHEN DOES A FUNCTION *NOT* SATISFY EITHER OF THE ABOVE THEOREMS?**

A. Sometimes you'll be asked to verify if a certain function satisfies either the MVT or Rolle's Theorem. All you need to do is to make sure it satisfies <u>all</u> of the hypotheses.

Example 1: $f(x) = \dfrac{1}{x}$ does not satisfy either theorem on an interval containing the origin because this function is not continuous (and hence, not differentiable) there.

Example 2: $g(x) = (x - 4)^{\frac{2}{3}}$ does not satisfy either theorem on any interval containing the point (4, 0) because, though continuous there, it is not differentiable there (that is, $g'(4)$ does not exist, $g(x)$ has a cusp there).

Example 3: The function $f(x) = x^2$ does not satisfy Rolle's Theorem on (0, 1) because $f(0) \neq f(1)$.

Keep in Mind...

➤ Don't confuse the Mean Value Theorem with Rolle's Theorem. Remember: for Rolle's Theorem you must set the function's derivative equal to zero, but for the Mean Value Theorem you must set the function's derivative equal to the slope of the secant between $x = a$ and $x = b$.

➤ Also, for both theorems, remember that the *c* value you are looking for is a number between *a* and *b*, but it cannot be *a* or *b*. If there is a *c* value that falls outside of the given interval, reject it.

➤ Rolle's Theorem applies to a function on [*a*, *b*] only if $f(a) = f(b)$.

CHAPTER 10
PRACTICE PROBLEMS

(See solutions on page 207)

1. Find the *c* value guaranteed by the Mean Value Theorem for
   $f(x) = \dfrac{1}{x-1}$ on [2, 4].

2. Find the *c* value(s) guaranteed by Rolle's Theorem for
   $y = 2\cos(3x)$ on $[-\pi, \pi]$.

3. Does $y = \ln(x)$ satisfy the Mean Value Theorem on [1, *e*]? If yes,
   find *c*. If not, explain why not.

4. Does $y = \ln(x)$ satisfy Rolle's Theorem on any interval? Explain.

5. If $f(x)$ is a polynomial function passing through the following
   points, what is the minimum number of roots it has on [0, 5]?

Newton's Method and Euler's Method

I. NEWTON'S METHOD*

A. The concept of derivative is used to find the roots of a function. The idea here is to find the equation of the tangent line repeatedly.

   1. Suppose that $f(x)$ is continuous on $[a, b]$ and differentiable on (a, b) and that $f(a)$ and $f(b)$ have different signs. Then, $f(x)$ must have at least one root, $x = x_0$, where $a < x_0 < b$. Choose an x-value in interval (a, b) and find the equation of the tangent line at this x-value, call it $x = x_1$. The x-intercept of this tangent line, call it $x = x_2$, is an approximation to one of the function's roots. Find the equation of the tangent line at $x = x_2$. The x-intercept of this tangent line, call it $x = x_3$, is a better approximation to the root. Repeat this process as many times as the problem asks, using the x-intercept as your new x-value every time. Since we are looking for the x-intercept of the tangent line, it is helpful to find an equation for it. The x-intercept is the value of x_1 that occurs when $y_1 = 0$ in the equation $y_1 - y_0 = m(x_1 - x_0)$. This yields: $-y_0 = mx_1 - mx_0$.

   Solving for x_1, we have $x_1 = x_0 - \dfrac{y_0}{m}$. More clearly,

   $x_1 = x_0 - \dfrac{f(x_0)}{f'(x_0)}$. In general, $x_{n+1} = x_n - \dfrac{f(x_n)}{f'(x_n)}$.

Test Tip

Certainly, if you do not want to memorize yet another formula, you can always just use the equation of the tangent line and find the x-intercept that way.

   2. In the graph of $f(x)$ that follows, you can see that the x values (which represent the x-intercepts of the tangent lines) approach the root of the function from right to left.

*Newton's Method is no longer a Calculus AB topic, but is recommended knowledge.

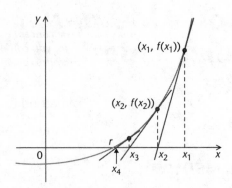

Example: find an approximation to the root of $f(x) =$ $x^2 - 2x - 1$ on $[2, 3]$ using two iterations of Newton's method. Note that $f(x)$ is a polynomial and thus continuous and differentiable at every point in its domain; also, $f(2) < 0$ and $f(3) > 0$ so $f(x)$ must have at least one root in $[2, 3]$. Choose $c_1 = 2.5$ and find x_1. $x_1 = x_0 - \dfrac{f(x_0)}{f'(x_0)}$

$\rightarrow x_1 = 2.5 - \dfrac{f(2.5)}{f'(2.5)} \rightarrow x_1 = 2.41666666$. Repeating the

process again, $x_2 = x_1 - \dfrac{f(x_1)}{f'(x_1)} \rightarrow x_2 = 2.417 - \dfrac{f(2.417)}{f'(2.417)} \rightarrow$

$x_2 = 2.41421630$. The root given by the calculator is $x = 2.4142136$.

II. EULER'S METHOD*

A. This method is used for approximating values of a function given a point on the function, the function's derivative, and the step size for x (the smaller the step size, the better the approximations).

B. Using the given point on a function, (x_0, y_0), the function's derivative, and the step size for x, Δx, one can approximate the y values of the function at $x_1 = x_0 + \Delta x$, $x_2 = x_1 + 2\Delta x$, $x_3 = x_2 + 3\Delta x$ and so on. Once again, starting with the equation of the tangent line to $f(x)$ at (x_0, y_0), we have: $y_1 - y_0 = m(x_1 - x_0) \rightarrow y_1 - y_0 = m\Delta x \rightarrow y_1 = y_0 + m\Delta x \rightarrow f(x_1) = f(x_0) + f'(x_0)\Delta x$. In general, $f(x_{n+1}) = f(x_n) + f'(x_n)\Delta x$.

*This is a BC Calculus topic.

Test Tip

If you thoroughly understand the concept and do not wish to memorize yet another formula, you can always use the equation of the tangent line to approximate the values of y.

Example: Let $f(x) = x^2$. Use Euler's method to approximate $f(2.6)$ given that $f(2) = 4$, $\dfrac{dy}{dx} = 2x$ and $\Delta x = 0.2$. We need to calculate $f(2.2)$ which will help us calculate $f(2.4)$ which will help us calculate $f(2.6)$. Here, $x_0 = 2$. So, $f(2.2) = f(2) + f'(2)\Delta x \rightarrow f(2.2) = 4 + 2(2)(0.2) \rightarrow f(2.2) = 4.8$. Repeating this process with $x_1 = 2.2$, we have: $f(2.4) = f(2.2) + f'(2.2)\Delta x \rightarrow f(2.4) = 4.8 + 2(2.2)(0.2) \rightarrow f(2.4) = 5.68$. One last iteration, with $x_2 = 2.4$: $f(2.6) = f(2.4) + f'(2.4)\Delta x \rightarrow f(2.6) = 5.68 + 2(2.4)(0.2) \rightarrow f(2.6) = 6.64$. The approximation becomes increasingly less accurate as x gets larger because the error gets larger with every iteration. Below is a graphical representation of the original function, $y = x^2$ and its approximation:

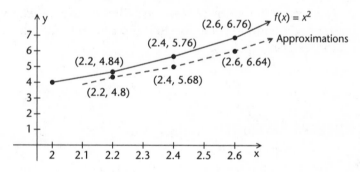

Keep in Mind...

➤ Do not round off your answers until the last step of the problem. And round off to at least 3 decimal places.

➤ Do only the number of iterations asked in the problem.

➤ Try to understand the concept behind these methods instead of just blindly memorizing the formulas—this goes for everything in this course!

➤ Sometimes you will not be asked explicitly to use Newton's method to solve a problem; in such cases, you must identify the need for it yourself. Look for phrases like "approximate the root" or something to that effect.

CHAPTER 11
PRACTICE PROBLEMS

(See solutions on page 208)

1. Approximate the root of $y = (5x - 3)^3$ on $[0, 1]$ using 6 iterations and $\Delta x = 0.5$. Find the actual answer and then find the error.

2. How many iterations of Newton's Method are necessary for approximating the solution to the problem in question 1 to three decimal places?

3. Given $f(3) = \dfrac{1}{4}$, $\dfrac{dy}{dx} = -\dfrac{1}{(x+1)^2}$, $\Delta x = 0.1$, evaluate $f(3.3)$.

PART IV
INTEGRALS

Types of Integrals, Interpretations and Properties of Definite Integrals, Theorems

I. TYPES OF INTEGRALS

A. *Indefinite integrals* have no limits, $\int f(x)dx$. This represents the antiderivative of $f(x)$. That is, if $\int f(x)dx = F(x) + C$, then

$F'(x) = f(x)$. When taking an antiderivative of a function, don't forget to add C! For instance, $\int 2xdx = x^2 + C$ (The constant C is necessary because the antiderivative of $f(x) = 2x$ could be $F(x) = x^2$ or $F(x) = x^2 + 1$ or $F(x) = x^2 - 2$, and so on.) Sometimes, you are given an initial condition that allows you to find the value of C. For instance, find the antiderivative, $F(x)$, of $f(x) = 2x$, given that $F(0) = 1$. Then,

$F(x) = \int 2xdx = x^2 + C \rightarrow F(0) = (0)^2 + C = 1 \rightarrow C = 1 \rightarrow F(x) = x^2 + 1.$

Another way of posing this question is: Find y if $\dfrac{dy}{dx} = 2x$ and

$y\big|_{x=0} = 1$. The equation $\dfrac{dy}{dx} = 2x$ is called a differential equation (more on this later) because it contains a derivative.

B. *Definite integrals* have limits $x = a$ and $x = b$, $\int_a^b f(x)dx$. If $f(x)$ is

continuous on $[a, b]$ and $F'(x) = f(x)$, then $\int_a^b f(x)dx = F(b) - F(a)$

(The First Fundamental Theorem of Calculus.)

1. A definite integral value could be positive, negative, zero or infinity. When used to find area, the definite integral must have a positive value.

i. If $f(x) > 0$ on $[a, b]$, then $\int_a^b f(x)dx > 0$ and geometrically it represents the area between the graph of $f(x)$ and the x-axis on the interval $[a, b]$. For example, $\int_0^3 2xdx = x^2\Big]_0^3 = 9$ square units. Note that this could also have been solved geometrically because the area in question is that of a right triangle with a base of 3 units and a height of 6 units.

$(A_\triangle = \dfrac{1}{2}bh = \dfrac{1}{2}3(6) = 9 \text{ units}^2)$

Solving an area problem geometrically is really helpful when the question involves the integral of a piecewise linear function, for instance, $\int_{-1}^3 |x|dx$. This represents the area between the function $f(x) = |x|$ and the x-axis between $x = -1$ and $x = 3$. Noticing that this area is that of two right triangles, we have: $\int_{-1}^3 |x|dx = \dfrac{1}{2}1(1) + \dfrac{1}{2}3(3) = 5 \text{ units}^2.$

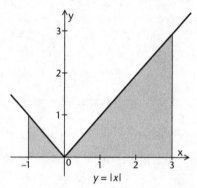

$y = |x|$

Remember that an absolute-value function is always made up of two pieces which you most often must consider separately because each piece is defined on a different interval. For instance, recall that $|x| = \begin{cases} x, & x \geq 0 \\ -x, & x < 0 \end{cases}$. So, algebraically,

$\int_{-1}^3 |x|dx = \int_{-1}^0 -xdx + \int_0^3 xdx = -\dfrac{x^2}{2}\Big]_{-1}^0 + \dfrac{x^2}{2}\Big]_0^3 = \left(0 + \dfrac{1}{2}\right) + \left(\dfrac{9}{2} - 0\right) =$

5 units^2. This would take much more time, especially when the functions get more complicated.

ii. If $f(x) < 0$ on $[a, b]$ then $\int_a^b f(x)dx < 0$ and the area between the graph of $f(x)$ and the x-axis on the interval $[a, b]$ is represented either by $\int_a^b |f(x)|dx$ or by $\left|\int_a^b f(x)dx\right|$. For instance, $\int_{-1}^1 (x^2 - 1)dx$. If you are simply asked to evaluate the integral, do so. That is,

$$\int_{-1}^1 (x^2 - 1)dx = \frac{x^3}{3} - x\Big]_{-1}^1 = \left(\frac{1}{3} - 1\right) - \left(\frac{-1}{3} - (-1)\right) = -\frac{4}{3}.$$

However, if the question asks for area, use the absolute value since area is always positive. That is, write either

$$\int_{-1}^1 |(x^2 - 1)|dx = -\int_{-1}^1 (x^2 - 1)dx = -\left(\frac{x^3}{3} - x\right)\Big]_{-1}^1 = -\left[\left(\frac{1}{3} - 1\right) - \left(\frac{-1}{3} - (-1)\right)\right] = \frac{4}{3} \text{ square units or, } \left|\int_{-1}^1 (x^2 - 1)dx\right| = \frac{x^3}{3} - x\Big]_{-1}^1 =$$

$$\left|\left(\frac{1}{3} - 1\right) - \left(\frac{-1}{3} - (-1)\right)\right| = \frac{4}{3} \text{ square units.}$$

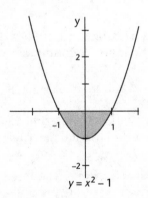

$y = x^2 - 1$

iii. If $f(x)$ is positive for some values of x and negative for other values of x on $[a, b]$, then the area between the graph of $f(x)$ and the x-axis on the interval $[a, b]$ is represented by $\int_a^b |f(x)|dx$. For example, the area between $f(x) = x^2 - 1$ and the x-axis on [0, 2] is given by

$$\int_0^2 |x^2 - 1| dx = \int_0^1 -(x^2 - 1) dx + \int_1^2 (x^2 - 1) dx = -\frac{x^3}{3} + x\Big]_0^1 +$$

$$\frac{x^3}{3} - x\Big]_1^2 = \left(-\frac{1}{3} + 1\right) + \left[\left(\frac{8}{3} - 2\right) - \left(\frac{1}{3} - 1\right)\right] = 2 \text{ units}^2. \text{ If you}$$

simply take the integral, $\int_0^2 (x^2 - 1) dx = \frac{x^3}{3} - x\Big]_0^2 = \frac{8}{3} - 2 = \frac{2}{3}$, the

answer does not represent the area between the function

and the x-axis, it represents the difference between the area above and the area below the x-axis.

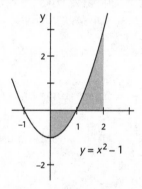

$y = x^2 - 1$

C.* *Improper Integrals* have one or both limits equal to either positive or negative infinity or are discontinuous on the given interval.

1. $\int_1^\infty \frac{1}{x^2} dx = \lim_{l \to \infty} \int_1^l \frac{1}{x^2} dx = \lim_{l \to \infty} \left(-\frac{1}{x}\right)\Big]_1^l = 1.$ If the answer is a

   constant, we say that the integral converges. If the answer is $\pm\infty$, we say that the integral diverges.

2. $\int_{-\infty}^\infty e^x dx = \int_{-\infty}^0 e^x dx + \int_0^\infty e^x dx = \infty.$ The integral diverges.

3. $\int_{-1}^1 \frac{1}{x^2} dx.$ This is an improper integral because $y = \frac{1}{x^2}$ is

   discontinuous within the interval [−1, 1] (at x = 0). Thus,

   $\int_{-1}^1 \frac{1}{x^2} dx = \int_{-1}^0 \frac{1}{x^2} dx + \int_0^1 \frac{1}{x^2} dx = \infty - \infty.$ This integral also diverges.

*This is a BC Calculus topic.

II. PROPERTIES OF DEFINITE INTEGRALS

1. $\displaystyle\int_a^b kf(x)dx = k\int_a^b f(x)dx$, for any constant k

2. $\displaystyle\int_a^a f(x)dx = 0$

3. $\displaystyle\int_a^b f(x)dx = -\int_b^a f(x)dx$

4. $\displaystyle\int_a^b f(x)dx + \int_b^c f(x)dx = \int_a^c f(x)dx$

5. $\displaystyle\int_a^b [f(x) \pm g(x)]dx = \int_a^b f(x)dx \pm \int_a^b g(x)dx$

6. If $f(x) \le g(x)$ on $[a, b]$, then $\displaystyle\int_a^b f(x)dx \le \int_a^b g(x)dx$

7. If $f(x)$ is even on $[-a, a]$, then $\displaystyle\int_{-a}^a f(x)dx = 2\int_0^a f(x)dx$

8. If $f(x)$ is odd on $[-a, a]$, then $\displaystyle\int_{-a}^a f(x)dx = 0$

9. $\displaystyle\int_a^b f'(x)dx = f(b) - f(a)$

III. THEOREMS

A. The First Fundamental Theorem of Calculus states that if $f(x)$ is continuous on $[a, b]$ and $F'(x) = f(x)$, then $\displaystyle\int_a^b f(x)dx = F(b) - F(a)$.

B. The Second Fundamental Theorem of Calculus states that if $f(x)$ is continuous on $[a, b]$, then $\displaystyle\frac{d}{dx}\int_a^x f(t)dt = f(x)$. In general,

$$\frac{d}{dx}\int_{h(x)}^{g(x)} f(t)dt = f(g(x))g'(x) - f(h(x))h'(x).$$

C. <u>The Mean Value Theorem for Integrals</u>—If $f(x)$ is continuous on

[a, b], then there is a c in [a, b], such that $\displaystyle\int_a^b f(x)\,dx = f(c)(b-a)$.

D. <u>Average Value of a Function</u>—If $f(x)$ is continuous on [a, b], the

average value of $f(x)$ on [a, b] is given by $A = \dfrac{1}{b-a}\displaystyle\int_a^b f(x)\,dx$ (this

can be derived from the Mean Value Theorem for Integrals).
This is not to be confused with the average rate of change of
a function, $f(x)$, on an interval, [a, b] which is the slope of the

secant of $f(x)$ on [a, b], (average rate of change = $\dfrac{f(b)-f(a)}{b-a}$),

and is used to approximate the slope of the tangent line at a
point inside [a, b].

<u>Note</u>: The average value of $f'(x)$ on [a, b] is equal to the
average rate of change of $f(x)$ on [a, b]. That is,

$$\frac{1}{b-a}\int_a^b f'(x)\,dx = \frac{f(b)-f(a)}{b-a}.$$

Keep in Mind...

➤ Don't forget to add the constant, C, when finding an indefinite
integral.

➤ Whenever possible, try to work backwards to find the
antiderivative of a function, it might save time.

➤ Don't confuse $\left|\int f(x)\,dx\right|$ with $\int |f(x)|\,dx$. They are the same only if
$f(x) > 0$.

➤ Remember that an absolute value function is two functions in one. Thus, when integrating an absolute value function, integrate each piece separately or do the problem graphically.

CHAPTER 12
PRACTICE PROBLEMS

(See solutions on page 209)

1. Find the area bounded by $y = 1 - x^2$ and the x-axis on $[0, 2]$.

2. Given $\int_a^b f(x)dx = 5$ and $\int_a^b g(x)dx = -3$ evaluate:

   (A) $\int_a^b [3f(x) - 2g(x)]dx$

   (B) $\int_b^a 6f(x)dx + \int_a^a \dfrac{g(x)}{\pi}dx$

3. (A) $\dfrac{d}{dx}\int_0^x \sqrt{3t+1}\,dt$

   (B) $\dfrac{d}{dx}\int_0^{4x^2} \dfrac{1}{e^t}\,dt$

4. Find the average value of $y = ex^3 + x$ on $[-1, 1]$.

5. Evaluate: $\int_4^\infty e^{-x}\,dx$

Riemann Sums (LRAM, RRAM, MRAM) and the Trapezoid Rule

I. **RIEMANN SUMS (LRAM, RRAM, MRAM)** are used to approximate the area between a function and the *x*-axis by slicing the area into thin vertical rectangles. (Riemann sums are sometimes used to approximate the area between a function/relation and the *y*-axis.)

A. LRAM—L̲eft R̲ectangle A̲pproximation M̲ethod. To approximate the area between a function, $f(x)$, and the *x*-axis on $[a, b]$, slice the area into vertical rectangular strips each of width Δx (the value of Δx will be given in the problem). Starting on the left, create rectangles and add up all their areas. For instance, below is the graph of $f(x)$ on $[a, b]$. To approximate this area using LRAM, create rectangles as shown below:

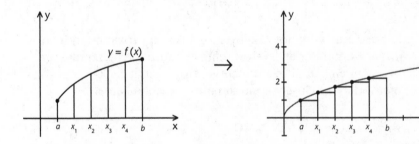

Then, the area between $f(x)$ and the *x*-axis on $[a, b]$ can be approximated by: $A \approx \Delta x(f(a) + f(x_1) + f(x_2) + f(x_3) + f(x_4))$.

This method is called the Left Rectangle Approximation Method because the upper left corner of each rectangle is on the curve. Note that in this case the approximation is an underestimation of the area since the area between the curve and the rectangles is left out.

B. RRAM—R̲ight R̲ectangle A̲pproximation M̲ethod. Using the function $f(x)$, seen on the prior page, create rectangles starting on the right such that the upper right corner of each rectangle is on the curve. This area then is represented by $A \approx \Delta x (f(x_1) + f(x_2) + f(x_3) + f(x_4) + f(b))$. Note that in this case the approximation is an *overestimation* of the actual area since the rectangles include more than just the area below the curve.

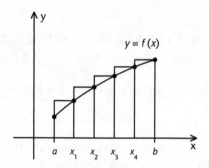

C. MRAM—M̲idpoint R̲ectangle A̲pproximation M̲ethod. Using the function $f(x)$, above, create rectangles such that the height of each rectangle is in the middle and the midpoint of the upper width of each rectangle is on the curve. This area is represented by:

$$A \approx \Delta x \left(f\left(\frac{a+x_1}{2}\right) + f\left(\frac{x_1+x_2}{2}\right) + f\left(\frac{x_2+x_3}{2}\right) + f\left(\frac{x_3+x_4}{2}\right) + f\left(\frac{x_4+b}{2}\right) \right).$$

In this case we're not exactly sure if this approximation is an underestimate or an overestimate because the rectangles are below as well as above the curve. However, it is clear that this method more closely approximates the actual area.

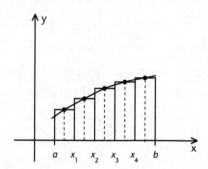

D. Trapezoid Rule—Given the function $f(x)$, seen earlier, connect the top endpoints of the vertical line segments, thus creating trapezoids. Add up the areas of all the trapezoids. This is an approximation of the area between the curve and the x-axis. Since the area of a trapezoid is $A = \dfrac{1}{2}h(b_1 + b_2)$,

the sum of the areas of the trapezoids below

is $A = \dfrac{1}{2}\Delta x \, (f(a) + 2f(x_1) + 2f(x_2) + 2f(x_3) + 2f(x_4) + f(b))$.

Note that the first base and the last base do not repeat, so they are not doubled. However, the inside bases are counted twice because adjacent trapezoids share a base. Also note that $h = \Delta x$.

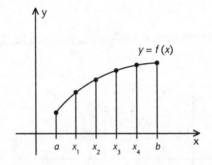

E. An example which illustrates all the above methods: approximate the area between the graph of $f(x) = \sqrt{x}$ and the x-axis on the interval [1, 4] using $\Delta x = 1$:

1. LRAM: $A \approx 1(\sqrt{1} + \sqrt{2} + \sqrt{3}) = 4.14626437$

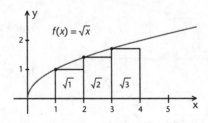

2. RRAM: $A \approx 1(\sqrt{2} + \sqrt{3} + \sqrt{4}) = 5.14626437$

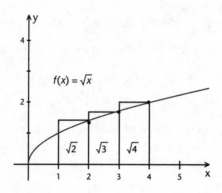

3. MRAM: $A \approx 1(\sqrt{1.5} + \sqrt{2.5} + \sqrt{3.5}) = 4.676712395$

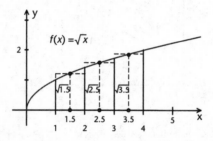

4. Trapezoid Rule: $A \approx \dfrac{1}{2}(1)(\sqrt{1} + 2\sqrt{2} + 2\sqrt{3} + \sqrt{4}) = 4.64626437$

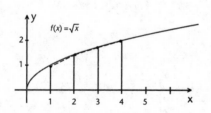

Trapezoid 1: $\dfrac{1}{2}(\sqrt{1} + \sqrt{2})$

Trapezoid 2: $\dfrac{1}{2}(\sqrt{2} + \sqrt{3})$

Trapezoid 3: $\dfrac{1}{2}(\sqrt{3} + \sqrt{4})$

5. The actual area: $A = \displaystyle\int_{1}^{4} \sqrt{x}\,dx = 4.\overline{6}$

Keep in Mind...

➤ Whichever estimation you use, if it calculates more than the given area, it is an overestimation. If it calculates less than the given area, it's an underestimation.

➤ When asked to calculate the area between a curve and an axis, draw the diagram; it always helps.

➤ Generally, the MRAM and TRAP methods give better approximations than the LRAM and RRAM methods.

CHAPTER 13
PRACTICE PROBLEMS

(See solutions on page 210)

1. Approximate the area between $y = \sqrt{x+4}$ and the x-axis from $x = -3$ to $x = 2$ using 5 equal subdivisions by using

   (A) LRAM

   (B) RRAM

   (C) MRAM

   (D) TRAP

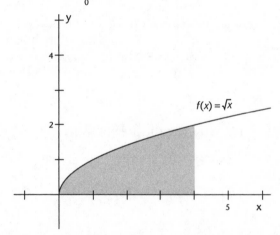

Applications of Antidifferentiation

I. AREA IN CARTESIAN COORDINATES

A. Area between a curve and the *x*-axis

   1. The area between $f(x)$ and the *x*-axis, if $f(x) \geq 0$ from $x = a$ to $x = b$, is represented by $\int_a^b f(x)dx$.

      i. For instance, the area between $f(x) = \sqrt{x}$ and the *x*-axis on [0, 4], is given by $\int_0^4 \sqrt{x}\,dx = 5.\overline{3}$.

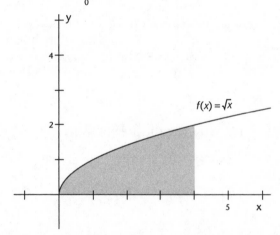

   2. The area between $f(x)$ and the *x*-axis, if $f(x) \leq 0$ from $x = a$ to $x = b$, is represented by $\left| \int_a^b f(x)dx \right|$.

i. For instance, the area between $f(x) = x^2 - 4$ on $[-2, 2]$, is

given by $\left| \int_{-2}^{2} (x^2 - 4) \, dx \right| = 10.\overline{6}$.

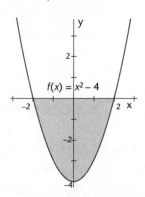

$f(x) = x^2 - 4$

3. The area between $f(x)$ and the x-axis, if $f(x)$ is sometimes negative and sometimes positive from $x = a$ to $x = b$, is

represented by $\int_{a}^{b} |f(x)| \, dx$. Evaluate the integral on the

interval(s) on which $f(x) \geq 0$, and add it to the absolute value of the integral on the intervals(s) on which $f(x) \leq 0$. That is, suppose that $f(x) \geq 0$ on $[a, b]$ and $f(x) < 0$ on $[c, d]$ where $a < b < c < d$. The area between $f(x)$ and the x-axis on $[a, d]$ is

given by $\int_{a}^{b} f(x) \, dx + \left| \int_{c}^{d} f(x) \, dx \right|$.

i. For instance, the area between $f(x) = 1 - x^2$ and the

x-axis on $[0, 3]$, is given by either $\int_{0}^{3} |1 - x^2| \, dx = 7.\overline{3}$ or,

equivalently, $\int_{0}^{1} (1 - x^2) \, dx + \left| \int_{1}^{3} (1 - x^2) \, dx \right| = 7.\overline{3}$.

4. To find the area between $f(x)$ and the y-axis on $y = c$ to $y = d$, rewrite the equation in terms of y first. That is, rewrite the equation in the form $x = g(y)$. If $g(y) > 0$ the area

is represented by $\int_{c}^{d} g(y) \, dy$. If $g(y) < 0$ the area is represented

by $\left| \int_{c}^{d} g(y) \, dy \right|$. If $g(y) > 0$ on $[a, b]$ and $g(y) < 0$ on $[c, d]$

the area between $g(y)$ and the y-axis on $[a, d]$ is given by

$$\int_a^b g(y)dy + \left|\int_c^d g(y)dy\right| \text{ or, equivalently, } \int_a^d |g(y)|dy.$$

i. For instance, the area between $f(x) = e^x$ and the y-axis

    from $y = 1$ to $y = 2$, is given by $\int_1^2 (\ln(y))dy = .3862943611.$

    Note that $f(x) = e^x \rightarrow y = e^x \rightarrow x = \ln(y)$.

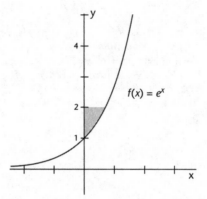

ii. Similarly, the area between $f(x) = e^x$ and the y-axis from

    $y = \dfrac{1}{2}$ to $y = 1$, is given by $\left|\int_{\frac{1}{2}}^{1} (\ln(y))dy\right| = .1534264097.$

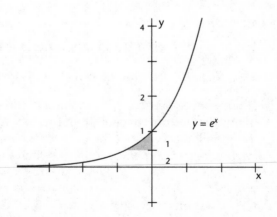

iii. Also, the area between $f(x) = e^x$ and the y-axis from
$y = \dfrac{1}{2}$ to $y = 2$, is given by

$$\left|\int_{\frac{1}{2}}^{1}(\ln(y))\,dy\right| + \int_{1}^{2}(\ln(y))\,dy = .5397207708 \text{ or}$$

$$\int_{\frac{1}{2}}^{2}|\ln(y)|\,dy = .5397267197$$

Test Tip

Note that although these two methods are equivalent mathematically, the calculator is using a variation of Riemann sum calculations and thus an approximation, but it will always be accurate to 3 decimal places.

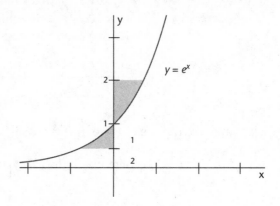

B. Area between two curves

1. The area between $f(x)$ and $g(x)$, where $g(x) \le f(x)$ from
$x = a$ to $x = b$, is represented by $\int_{a}^{b}(f(x) - g(x))\,dx$.

Loosely speaking, this is the integral of the top function minus the bottom function. If the answer is negative, then this is an indication that the order of the functions in the integrand is wrong and you must switch the functions around.

i. For instance, the area between $f(x) = x$ and $g(x) = x^2$ on
[2, 3] is given by $\int_{2}^{3}(x^2 - x)\,dx = 3.8\overline{3}$.

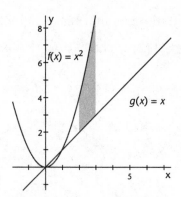

2. The area between $f(y)$ and $g(y)$, where $g(y) \le f(y)$ from $y = c$

to $y = d$, is represented by $\int_{c}^{d}(f(y) - g(y))dy$. Loosely speaking,

this is the integral of the right function minus the left function. If the answer is negative, then this is an indication that the order of the functions in the integrand is wrong and you must switch the functions around.

i. For instance, the area between $f(y) = y$ and $g(y) = e^y$ on

$1 \le y \le 2$, is given by $\int_{1}^{2}(e^y - y)dy = 3.17077427$

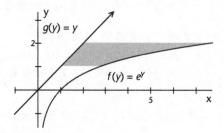

II. AREA IN POLAR COORDINATES*

A. Area inside a polar curve

1. The area inside a polar curve, $r = f(\theta)$, is represented by

$$\frac{1}{2}\int_{\theta_1}^{\theta_2}(f(\theta))^2 d\theta.$$

*This is a BC Calculus topic.

i. For example, the area of cardioid $r = 2 + 2\sin(\theta)$, is given by $\dfrac{1}{2}\displaystyle\int_0^{2\pi}(2 + 2\sin(\theta))^2\,d\theta = 18.84955592$

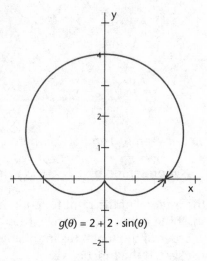

$g(\theta) = 2 + 2 \cdot \sin(\theta)$

B. Area between two polar curves

1. The area between $h(\omega)$ and $g(\theta)$, if $h(\theta) > g(\theta)$ is represented by $\dfrac{1}{2}\displaystyle\int_{\theta_1}^{\theta_2}[(h(\theta))^2 - (g(\theta))^2]\,d\theta$ where the limits, θ_1 and θ_2, represent the angles at which the two curves intersect.

i. For example, the area between $h(\omega) = 1 + \cos(\omega)$ and $g(\omega) = 3\cos(\omega)$ in the first and fourth quadrants, is given by $\dfrac{1}{2}\displaystyle\int_{-\frac{\pi}{3}}^{\frac{\pi}{3}}[(3\cos(\theta))^2 - (1 + \cos(\theta))^2]\,d\theta$, where $\theta = \pm\dfrac{\pi}{3}$ are the angles at which the curves intersect.

Because of x-axis symmetry, an equivalent solution is:

$\displaystyle\int_0^{\frac{\pi}{3}}[(3\cos(\theta))^2 - (1 + \cos(\theta))^2]\,d\theta$, that is, double the area from $\theta = 0$ to $\theta = \dfrac{\pi}{3}$.

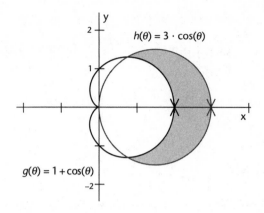

$h(\theta) = 3 \cdot \cos(\theta)$

$g(\theta) = 1 + \cos(\theta)$

III. LENGTH OF CURVE*

A. Length of curve in Cartesian coordinates

1. The length (also known as arc length) of a smooth Cartesian curve, $f(x)$, from $x = x_1$ to $x = x_2$, is represented by $L = \int_{x_1}^{x_2} \sqrt{1 + \left(\dfrac{dy}{dx}\right)^2} \, dx$

   i. For example, the length of $f(x) = x^2$ on $[0, 1]$ is given by
   $$L = \int_0^1 \sqrt{1 + (2x)^2} \, dx = 1.478942858$$

2. The length of a smooth Cartesian curve, $f(y)$, from $y = y_1$ to $y = y_2$ is represented by $L = \int_{y_1}^{y_2} \sqrt{1 + \left(\dfrac{dx}{dy}\right)^2} \, dy$

   i. For example, the length of $f(y) = e^y + y$ from $y_1 = 2$ to $y_2 = 3$ is given by $L = \int_2^3 \sqrt{1 + (e^y + 1)^2} \, dy = 13.73557854$

B. Length of curve in parametric form

1. The length of a parametric curve, from $t = t_1$ to $t = t_2$ is represented by $L = \int_{t_1}^{t_2} \sqrt{\left(\dfrac{dx}{dt}\right)^2 + \left(\dfrac{dy}{dt}\right)^2} \, dt$

*This is a BC Calculus topic.

i. For example, the length of the parametric curve represented by $\begin{cases} x(t) = t \\ y(t) = t^2 \end{cases}$ from $t = 1$ to $t = 3$ is given by

$$L = \int_1^3 \sqrt{(1)^2 + (2t)^2}\, dt = 8.26814590\ldots$$

IV. VOLUME

A. *Washer Method*—The washer method is used when the cross sections of the solid are washers (generally when the volume is that of a solid which has been created by rotating a region bounded by <u>two</u> curves about an axis).

1. If rotating the region between $f(x)$ and $g(x)$, where $f(x) \geq g(x)$, <u>about the x-axis</u>, the volume of the resulting solid is represented by $\pi \int_{x_1}^{x_2} [(f(x))^2 - (g(x))^2]\, dx$ where x_1 and x_2 represent the x-values of the intersection points of the two functions. Loosely speaking, this is the integral of the top function squared minus the bottom function squared. Don't forget the π!

   i. For example, the volume of the solid formed when the region between $f(x) = 3$ and $g(x) = x^2$ is revolved about the x-axis, is given by $\pi \int_{-\sqrt{3}}^{\sqrt{3}} [(3)^2 - (x^2)^2]\, dx = 78.35613253$.

   To find the x-values of the intersection points, set the functions equal to each other and solve for x. In this case, $3 = x^2 \rightarrow x = \pm\sqrt{3}$.

2. If rotating the region between $f(x)$ and $g(x)$, where $f(x) \geq g(x)$, about a <u>horizontal line, $y = k$</u>, the volume of the resulting solid is represented by $\pi \int_{x_1}^{x_2} [(f(x) - k)^2 - (g(x) - k)^2]\, dx$

   Don't forget the π!

i. For example, the volume of the solid formed when the region between $y = x^2 + 4$ and $y = x + 4$ is revolved about the line $y = 1$, is given by

$$\pi\int_0^1 [(x+4-1)^2 - (x^2+4-1)^2]\,dx = 3.560471674.$$

3. If rotating the region between $f(y)$ and $g(y)$, where $f(y) \geq g(y)$, <u>about the y-axis</u>, the volume of the resulting solid

is represented by $\pi\int_{y_1}^{y_2} [(f(y))^2 - (g(y))^2]\,dy$. Loosely speaking,

this is the integral of the right function squared minus the left function squared. Don't forget the π!

i. For instance, the volume of the solid formed by revolving the region bounded by $x = 3$ and $x = \ln(y)$ from $y = 1$ to $y = 3$ about the y-axis, is given by

$$\pi\int_1^3 [(3)^2 - (\ln(y))^2]\,dy = 53.31542496.$$

4. When rotating the region between $f(y)$ and $g(y)$, where $f(y) \geq g(y)$, about a <u>vertical line, $x = k$</u>, the volume of the

resulting solid is represented by $\pi\int_{y_1}^{y_2} [(f(y)-k)^2 - (g(y)-k)^2]\,dy$

or, equivalently, $\pi\int_{y_1}^{y_2} [(k-f(y))^2 - (k-g(y))^2]\,dy$, where y_1 and

y_2 represent the y-values of the intersection points of the two functions. Don't forget the π!

i. For example, the volume of the solid obtained by rotating $x = 3$ and $x = \ln(y)$ on $1 \leq y \leq 2$ about the line $x = 4$, is

given by $\pi\int_1^2 [(4-\ln(y))^2 - (4-3)^2]\,dy = 38.00686985.$

B. *Disk Method*—The disk method is used when the cross sections of the solid are disks. This method generally involves only one function. This is a simple case of the washer method in which $g(x) = 0$ or $g(y) = 0$. For instance, the volume of the solid formed by revolving $y = x^3$ about the x-axis on [0, 2] is given by

$$\pi\int_0^2 [(x^3)^2 - (0)^2]\,dx = \pi\int_0^2 [(x^3)^2]\,dx = 57.44626567.$$

C. *Cylindrical Shells Method**

1. Cylindrical shells provide an alternate method of finding volume which is particularly useful when rotating areas about vertical lines. One example is the volume of the solid formed by revolving the region bounded by the graph of $y = \sin(x)$ and the x-axis on $[0, \pi]$ about the y-axis. The idea behind this method is that the solid in question is sliced into concentric cylinders. The sum of the volumes of these cylindrical slices (not the volume inside the slices, but of the slices themselves) gives the volume of the solid. The volume of such a slice is given by $V_{cylinder} = 2\pi rh(thickness)$. So, generally speaking,

finding the volume of a solid using cylindrical shells yields

$2\pi\int_{a}^{b} (radius)(height)(thickness)$. When the cylindrical shells are

parallel to the y-axis, the thickness $= dx$.
When the cylindrical shells are parallel to the x-axis, the thickness $= dy$. The advantage to the shell method is that when rotating about the y-axis, we can keep our variables in terms of x.

i. For example, the volume of the solid formed by revolving the region bounded by the graph of $y = \sin(x)$ and the x-axis on $[0, \pi]$ about the y-axis, is given by

$2\pi\int_{0}^{\pi} (x)(\sin(x))\,dx = 19.7392088$.

ii. Similarly, the volume of the solid formed by revolving the region bounded by the graph of $x = (y-2)^2$ and $x = y$ about the line $y = 1$ is given by

$2\pi\int_{1}^{4} (y-1)(y-(y-2)^2)\,dy = 42.41150082$. Note that the

limits of the integral are the y-values of the intersection points of the two equations.

*This is no longer a BC topic, but remains recommended knowledge for both Calculus AB and BC.

V. DIFFERENTIAL EQUATIONS AND SLOPE FIELDS

A. *Differential equations* are equations that contain at least one derivative. To solve a differential equation means to find the original function. That is, if given $f'(x)$, you must find $f(x)$. This implies taking the antiderivative of $f'(x)$.

1. To solve a differential equation, separate and integrate. That is, algebraically manipulate the equation such that all the x terms are on one side of the equation and all the y terms are on the other. Then integrate each side. Each side will yield a constant of integration but combining them yields only one.

   i. For example, solve $\dfrac{dy}{dx} = \dfrac{1}{y}$ given that $y\,|_{x=1} = 3$.

   Separating the variables yields $y\,dy = dx$. Taking the antiderivative of both sides, $\int y\,dy = \int dx \;\rightarrow\; \dfrac{y^2}{2} + C_1 = x + C_2$. The two constants of integration get combined to yield $\dfrac{y^2}{2} = x + C_2 - C_1$ or simply, $\dfrac{y^2}{2} = x + C$. There is no need to write the two constants; writing only one C is acceptable. To find the value of C apply the initial condition, $y\,|_{x=1} = 3$.

   This yields: $\dfrac{3^2}{2} = 1 + C \;\rightarrow\; C = \dfrac{7}{2} \;\rightarrow\; \dfrac{y^2}{2} = x + \dfrac{7}{2}$. The equation may be left as is or it may be algebraically manipulated into one of the various other forms: $y^2 = 2x + 7$, or $y^2 - 2x = 7$, $y = \sqrt{2x + 7}$, but not $y = \pm\sqrt{2x + 7}$ as the point (1, 3) doesn't pass through $y = -\sqrt{2x + 7}$.

B. Slope fields are fields of slopes, literally. Given a family of differentiable functions, $y = f(x) + C$, imagine drawing a tiny tangent line at each point on these functions. The set of all of these tangent lines forms the slope field for the function. When given a differential equation, the original function can be obtained by drawing the slope field.

1. For example, given $\dfrac{dy}{dx} = -\dfrac{x}{y}$ and a point on the original function, $(\sqrt{2}, \sqrt{2})$. Substituting some x and y values into $\dfrac{dy}{dx} = -\dfrac{x}{y}$ helps us create the slope field, below.

x	y	$m = \dfrac{dy}{dx} = -\dfrac{x}{y}$
0	1	0
0	2	0
0	−1	0
0	−2	0
1	0	$-\infty$
1	1	−1
1	2	$-\dfrac{1}{2}$
1	−1	1
1	−2	$\dfrac{1}{2}$
−1	0	∞
−1	1	1
−1	2	$\dfrac{1}{2}$
−1	−1	−1
−1	−2	$-\dfrac{1}{2}$
2	0	$-\infty$
2	1	−2
2	−1	2
−2	0	∞
−2	1	2
−2	−1	−2

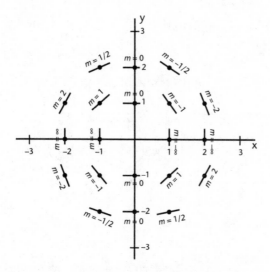

This slope field suggests that the function whose derivative is given belongs to the family of circles with the center at the origin.

Solving the differential equation by separating and integrating, yields

$$\frac{dy}{dx} = -\frac{x}{y} \rightarrow ydy = -xdx \rightarrow \int ydy = -\int xdx \rightarrow \frac{y^2}{2} = -\frac{x}{2} +$$

$$C \xrightarrow{\text{substitute in } (\sqrt{2},\sqrt{2}) \text{ to find } C} \frac{\left(\sqrt{2}\right)^2}{2} = -\frac{\left(\sqrt{2}\right)^2}{2} + C \rightarrow C = 2.$$

The particular solution to the given differential equation is

$\frac{y^2}{2} = -\frac{x^2}{2} + 2,$ or equivalently, $x^2 + y^2 = 4$, which represents a

circle with the center at the origin and radius 2 units.

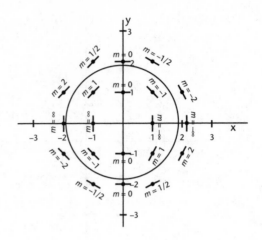

VI. MOTION

A. *Rectilinear motion in the Cartesian system.* Since the acceleration, velocity, and displacement of an object can be expressed as derivatives, that is, $v(t) = \dfrac{ds}{dt}$ and $a(t) = \dfrac{dv}{dt} = \dfrac{d^2s}{dt^2}$, it follows that they may also be expressed as antiderivatives.

1. The total *displacement* of an object moving in a straight line from $t = t_1$ to $t = t_2$ is represented by $\displaystyle\int_{t_1}^{t_2} v(t)dt$.

   The displacement equation is given by $\int v(t)dt$ and requires an initial condition to be given. Loosely speaking, the displacement equals the integral of velocity.

   i. For example, an object travels with velocity given by $v(t) = 2t^2 - 3t$ and it is given that $s(0) = 1$. Its displacement equation is given by $s(t) = \int (2t^2 - 2t)\,dt = \dfrac{2t^3}{3} - t^2 + C$. Since $s(0) = 1$, $1 = \dfrac{2(0)^3}{3} - 0^2 + C \;\rightarrow\; 1 = C \;\rightarrow\; s(t) = \dfrac{2t^3}{3} - t^2 + 1$.

    ii. The total displacement of the object in part i for the first two seconds is given by

$$s(t) = \int_0^2 (2t^2 - 2t)\,dt = \frac{2t^3}{3} - t^2 \Bigg]_0^2 = 1.\overline{3}$$

2. The total distance traveled by an object on the interval (t_1, t_2) is represented by $\int_{t_1}^{t_2} |v(t)|\,dt$. Loosely speaking, the total distance equals the integral of the speed.

    i. For example, the total distance traveled by the object in part 1ii, above, is given by $d = \int_0^2 |(2t^2 - 2t)|\,dt = 2$.

B. *Motion on a parametric/vector curve**

    1. In vector form, recall that the velocity of an object is given by $v = \langle x'(t), y'(t) \rangle$. Thus, the speed is given by

$$|v| = \sqrt{(x'(t))^2 + (y'(t))^2}\,dt,$$ also called the magnitude of the velocity vector. This also represents the speed in parametric form. Hence, the total distance traveled on such a curve is given by $d = \int_{t_1}^{t_2} \sqrt{(x'(t))^2 + (y'(t))^2}\,dt$ for both a parametric

curve, $\begin{cases} x = x(t) \\ y = y(t) \end{cases}$, and a vector curve $\langle x(t), y(t) \rangle$. Notice that

$d = \int_{t_1}^{t_2} \sqrt{(x'(t))^2 + (y'(t))^2}\,dt$ also represents the length of the

curve $\langle x(t), y(t) \rangle$ or $\begin{cases} x = x(t) \\ y = y(t) \end{cases}$ on (t_1, t_2).

    i. For example, the total distance traveled along $\begin{cases} x = 2t \\ y = \ln(t) \end{cases}$ on

$1 \le t \le 2$, is given by $d = \int_1^2 \sqrt{(2)^2 + \left(\frac{1}{t}\right)^2}\,dt = 2.120783012$.

This is a BC Calculus topic.

VII. EXPONENTIAL GROWTH AND DECAY AND LOGISTIC GROWTH*

A. *Exponential growth and decay* refer to the change in a quantity over time.

   1. The relationship between this quantity (bacteria, population, money in a bank account, etc.) and time is represented by the differential equation $\frac{dy}{dt} = ky$ (this states that the change in the quantity over time is proportional to the amount present at any time, t) whose solution is $y = y_0 e^{kt}$ where y represents the quantity at time t, y_0 represents the initial quantity, and k is a constant. If $k > 0$ then these equations represent exponential growth; if $k < 0$ these equations represent exponential decay.

Test Tip

> *The words and phrases "quantity increases exponentially," "quantity decreases exponentially," "change in quantity is proportional to the amount present," "exponential growth," "exponential decay," or simply "bacteria growth" indicate that you must use $y = y_0 e^{kt}$.*

   i. For example, originally there are 10 bacteria in a dish. Four hours later there are 15 bacteria in the dish. How long will it take for the number of bacteria to reach 30? Since the problem involves bacteria, it must fall into the category of exponential growth. Here, $y = 15$, $y_0 = 10$, $t = ?$ when $y = 30$. Hence, $y = y_0 e^{kt} \rightarrow 15 = 10e^{k(4)} \rightarrow 1.5 = e^{4k} \rightarrow \ln(1.5) = 4k \rightarrow$ $k = \dfrac{\ln(1.5)}{4}$. Then, $30 = 10e^{\frac{\ln(1.5)}{4}t} \rightarrow 3 = e^{\ln(1.5)\frac{t}{4}} \rightarrow$

   $3 = 1.5^{\frac{t}{4}} \rightarrow \ln(3) = \dfrac{t}{4}\ln(1.5) \rightarrow t = 10.83804517$ hours.

B.* *Logistic Growth* is growth that occurs when there are limiting factors present.

   1. For instance, the population of fish in a fish tank is growing logistically (as opposed to exponentially which implies no bound) because there are factors that slow the growth down such as the competition for space, food, and oxygen. At some point, the population will reach a maximum.

*This is a BC Calculus topic.

Test Tip

To recognize questions involving logistic growth, look for the terms "logistic growth" or spot the equations involved:

$$\frac{dP}{dt} = \frac{k}{M}P(M-P) \text{ or its solution, } P(t) = \frac{M}{1+Ae^{-kt}}, \text{ where}$$

P is the size of the population at any time t, k and A are constants, and M is the maximum population (also called the carrying capacity).

Some facts about logistic growth:

(a) The population grows fastest at $P = \dfrac{M}{2}$ (this is where

P(t) has an inflection point which is the maximum of $\dfrac{dP}{dt}$)

(b) $\lim\limits_{t\to\infty} P(t) = M$ (y = M is also the horizontal asymptote of P(t))

(c) The shape of the graph of $P(t) = \dfrac{M}{1+Ae^{-kt}}$:

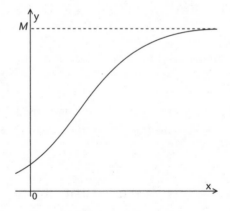

Keep in Mind...

➤ When finding volume, don't forget to include π. When entering the expression in the calculator, start with π so that you do not forget it.

➤ Memorize the component parts of the exponential growth/ decay formula and to the logistic growth formula, there will be no time for you to derive these.

➤ For cylindrical shells remember that when the shells are parallel to the *x*-axis, the thickness is *dy* and when the shells are parallel to the *y*-axis, the thickness is *dx*.

CHAPTER 14
PRACTICE PROBLEMS

(See solutions on page 212)

1. Find the area inside $r = 2 \sin(\theta)$ and check your answer using the formula for the area of the graph.

2. Find the area outside $r = 2 + 2 \sin(\theta)$ and inside $r = \cos(\theta)$.

3. Find the length of the curve represented by $\begin{cases} x(t) = e^t + 1 \\ y(t) = 2t + 1 \end{cases}$ on $0 \le t \le 1$.

4. Find the volume that results when the area between the graph of $y = \cos(x)$ and the *x*-axis from $x = 0$ to $x = \dfrac{\pi}{2}$ is revolved about the line $x = -1$.

5. Solve: $\dfrac{dy}{dx} = 2x(y + 1)$ given that $y(0) = 1$.

6. The growth of a given population is represented by $\dfrac{dP}{dt} = \dfrac{3}{10} P(10 - P)$ where *P* represents the population in millions.

   (A) When does the population grow fastest?
   (B) Evaluate: $\lim\limits_{t \to \infty} P(t)$ and explain the meaning of this answer.

Techniques of Integration

I. INTEGRATION TECHNIQUES

A. *U-substitution* is used to rewrite the integrand so that it is easily integrable. This method is used when the integrand is of the form $f(g(x))g'(x)$ where $g'(x)$ can be off by a constant factor. This is the opposite of the chain rule for derivatives.

1. To use this method for $\int f(g(x))g'(x)dx$, let $u = g(x)$. Then, $du = g'(x)dx$ and $\int f(g(x))g'(x)dx = \int f(u)du$.

   i. For instance, for $\int (6x - 2)\sqrt{3x^2 - 2x + 5}dx$, let $u = 3x^2 - 2x + 5$, $du = (6x - 2)dx$ and $\int (6x - 2)\sqrt{3x^2 - 2x + 5}dx = \int \sqrt{u}du = \frac{2}{3}u^{\frac{3}{2}} + C = \frac{2}{3}(3x^2 - 2x + 5)^{\frac{3}{2}} + C.$

Test Tip *Always rewrite the problem using the original variable unless otherwise directed.*

   ii. For $\int \frac{3x}{e^{5x^2}}dx$ let $u = 5x^2$ and $du = 10xdx \rightarrow \frac{1}{10}du = xdx$. So,

   $$\int \frac{3x}{e^{5x^2}}dx = 3\int \frac{x}{e^{5x^2}}dx = 3\int \frac{xdx}{e^{5x^2}} = 3\int \frac{\frac{1}{10}du}{e^u} = \frac{3}{10}\int \frac{1}{e^u}du =$$

   $$\frac{3}{10}\int e^{-u}du = -\frac{3}{10}e^{-u} + C = -\frac{3}{10}e^{-(5x^2)} + C.$$

B. *Powers of Trigonometric Functions**

1. $\int \sin^m(x)\cos^n(x)dx$	Method	Identities Used
If *m* and *n* are even	– Use the relevant identities to reduce the powers.	$\cos^2(x) = \dfrac{1}{2}(1+\cos(2x))$ $\sin^2(x) = \dfrac{1}{2}(1-\cos(2x))$
If *m* is odd	– Factor out a factor of $\sin(x)$. – Apply the relevant identity. – Let $u = \cos(x)$.	$\sin^2(x) = 1-\cos^2(x)$
If *n* is odd	– Factor out a factor of $\cos(x)$. – Apply the relevant identity. – Let $u = \sin(x)$.	$\cos^2(x) = 1-\sin^2(x)$

i. For example, $\int \sin^4(x)\cos^5(x)dx = \int \sin^4(x)\cos^4(x)\cos(x)dx =$

$\int \sin^4(x)(1-\sin^2(x))^2 \cos(x)dx$. Let $u = \sin(x)$. Then

$du = \cos(x)dx$ and $\int \sin^4(x)(1-\sin^2(x))^2 \cos(x)dx =$

$\int u^4(1-u^2)^2 du = \int(u^4 - 2u^6 + u^8)du = \dfrac{u^5}{5} - \dfrac{2u^7}{7} + \dfrac{u^9}{9} + C.$

Rewriting it in terms of x, $\int \sin^4(x)(1-\sin^2(x))^2 \cos(x)dx =$

$\dfrac{\sin^5(x)}{5} - \dfrac{2\sin^7(x)}{7} + \dfrac{\sin^9(x)}{9} + C.$

* This is not a current AP Calculus topic, but remains recommended knowledge for the Calculus BC exam.

2. $\int \tan^m(x)\sec^n(x)dx$	Method	Identities Used
If m is even and n is odd	– Use the relevant identities to reduce to secant powers only. – Use the reduction formula for powers of sec(x) (that is, $\int \sec^n(x)dx =$ $\dfrac{\sec^{n-2}(x)\tan x}{n-1} +$ $\dfrac{n-2}{n-1}\int \sec^{n-2}(x)dx$).	$\tan^2(x) = \sec^2(x) - 1$
If n is even	– Factor out a factor of $\sec^2(x)$. – Apply the relevant identity. – Let $u = \tan(x)$.	$\sec^2(x) = \tan^2(x) + 1$
If m is odd	– Factor out a factor of $\sec(x)\tan(x)$. – Apply the relevant identity. – Let $u = \sec(x)$.	$\tan^2(x) = \sec^2(x) - 1$

i. $\int \tan^2(x)\sec^4(x)dx = \int \tan^2(x)\sec^2(x)\sec^2(x)dx =$

$\int \tan^2(x)(\tan^2(x)+1)\sec^2(x)dx$. Let $u = \tan(x)$, then

$du = \sec^2(x)dx$ and $\int \tan^2(x)(\tan^2(x)+1)\sec^2(x)dx =$

$\int u^2(u^2+1)du = \int (u^4 + u^2)du = \dfrac{u^5}{5} + \dfrac{u^3}{3} + C$. Rewriting the

answer in terms of x, $\int \tan^2(x)\sec^4(x)dx = \dfrac{u^5}{5} + \dfrac{u^3}{3} + C =$

$\dfrac{\tan^5(x)}{5} + \dfrac{\tan^3(x)}{3} + C$.

C.* *Integration by parts* is used when the integrand is a product of unrelated functions, of the form $f(x)g'(x)$. Let $u = f(x)$ and $v = g(x)$. Then, $\int u\,dv = uv - \int v\,du$. This is the opposite of the product rule for derivatives.

Test Tip

There are problems in which this method might be used more than once. To decide which of the two functions to let equal u and which to let equal dv can be tricky, but you'll know when you've gone down the wrong path. Instead of becoming simpler, the problem becomes more difficult.

i. For example, $\int xe^x dx$. Let $u = x$ and $dv = e^x dx$. Then, $du = dx$ and $v = e^x$. Thus, $\int xe^x dx = xe^x - \int e^x dx = xe^x - e^x + C$.

ii. When solving $\int x^2 \sin(x)dx$, let $u = x^2$ and $dv = \sin(x)dx$. Then, $du = 2x\,dx$ and $v = -\cos(x)$. Thus, $\int x^2 \sin(x)dx = -x^2 \cos(x) + 2\int x\cos(x)dx$. Here we need to integrate by parts again. So, to integrate $\int x\cos(x)dx$ let $u = x$ and $dv = \cos(x)dx$. Then $du = dx$ and $v = \sin(x)$. $\int x\cos(x)dx = x\sin(x) - \int \sin(x)dx = x\sin(x) + \cos(x)$. We'll add the constant of integration in the next step.

So, $\int x^2 \sin(x)dx = -x^2 \cos(x) + 2\int x\cos(x)dx = -x^2 \cos(x) +$

$2[x\sin(x) + \cos(x)] + C = -x^2 \cos(x) + 2x\sin(x) + 2\cos(x) + C$.
In this case we used integration by parts twice.

iii. To solve $\int e^x \sin(x)dx$, let $u = e^x$ and $dv = \sin(x)dx$. Then, $du = e^x dx$ and $v = -\cos(x)$. So, $\int e^x \sin(x)dx =$

* This is a BC Calculus topic.

$-e^x \cos(x) + \int e^x \cos(x)dx$. We must use integration

by parts once more for $\int e^x \cos(x)dx$. Let $u = e^x$

and $dv = \cos(x)dx$. Then $du = e^x dx$ and $v = \sin(x)$.

$\int e^x \cos(x)dx = e^x \sin(x) - \int e^x \sin(x)dx$. Finally, the original

problem, $\int e^x \sin(x)dx = -e^x \cos(x) + \int e^x \cos(x)dx =$

$-e^x \cos(x) + e^x \sin(x) - \int e^x \sin(x)dx$. More

simply, $\int e^x \sin(x)dx = -e^x \cos(x) + e^x \sin(x) - \int e^x \sin(x)dx$.

Adding $\int e^x \sin(x)dx$ to both sides of the equation yields

$2\int e^x \sin(x)dx = -e^x \cos(x) + e^x \sin(x) \to \int e^x \sin(x)dx =$

$\dfrac{-e^x \cos(x) + e^x \sin(x)}{2} + C.$

D.* *Integration by partial fractions* is used by separating a fraction into
partial fractions. For instance, the partial fractions for

$\dfrac{1}{(x+1)(x-1)}$ are $\dfrac{-\dfrac{1}{2}}{(x+1)}$ and $\dfrac{\dfrac{1}{2}}{(x-1)}$ because $\dfrac{-\dfrac{1}{2}}{(x+1)} + \dfrac{\dfrac{1}{2}}{(x-1)} =$

$\dfrac{1}{(x+1)(x-1)}$. This helps when integrating $\int \dfrac{1}{x^2-1}dx =$

$\int \dfrac{1}{(x+1)(x-1)}dx = \dfrac{1}{2}\int\left(\dfrac{1}{x-1} - \dfrac{1}{x+1}\right)dx = \dfrac{1}{2}(\ln|x-1| - \ln|x+1|) +$

$C = \dfrac{1}{2}\left(\ln\left|\dfrac{x-1}{x+1}\right|\right) + C.$ To decompose $\dfrac{1}{(x+1)(x-1)}$ into its partial

fractions, let $\dfrac{1}{(x+1)(x-1)} = \dfrac{A}{x+1} + \dfrac{B}{x-1} \to \dfrac{1}{(x+1)(x-1)} =$

$\dfrac{x(A+B) + B - A}{(x+1)(x-1)} \to 1 = x(A+B) + B - A \to \begin{cases} 1 = B - A \\ 0 = A + B \end{cases} \to A = -\dfrac{1}{2}, B = \dfrac{1}{2}.$

Use this method when the integrand is a fraction with linear
factors in the denominator and the *u*-substitution cannot be used.

* This is a BC Calculus topic.

Keep in Mind...

➤ Remember that $\int \frac{1}{1-x} dx \neq \ln|1-x|+C$ but $\int \frac{1}{1-x} dx =$ $-\ln|1-x|+C$. This is often part of the method of partial fractions.

➤ When doing integration by parts, let dv be equal to the factor which is simpler to integrate. If you are not sure which factor to let equal u and which to let equal dv, and the integration becomes more cumbersome instead of simpler, then you've picked the factors wrong. Switch them.

➤ When deciding which integration method to use, eliminate the possibilities in order from easiest to most difficult—working backwards: u-substitution, integration by parts (generally used for a product of functions), integration by partial fractions (generally used for rational functions in which the denominator is a linear or quadratic function).

CHAPTER 15
PRACTICE PROBLEMS

(See solutions on page 214)

Integrate and state the method used:

1. $\int e^x \sqrt{e^x} dx$

2. $\int \sin^3(4x)\cos(4x)dx$

3. $\int x \ln x dx$

4. $\int \frac{2x}{x^2 - 3x + 2} dx$

5. $\int \frac{2}{3 + x^2} dx$

PART V

SEQUENCES AND SERIES

Sequences and Series

I. **SEQUENCES**—a sequence is a list of numbers separated by commas $a_1, a_2, a_3, \ldots, a_k, \ldots$, that may or may not have a pattern.

A. Arithmetic and geometric sequences

1. The formula for the nth term of an arithmetic sequence (one that is formed by adding the same constant repeatedly to an initial value) is $a_n = a_1 + (n-1)d$ where a_1 is the first term of the sequence, n is the number of terms in the sequence, and d is the common difference. The formula for the nth term of a geometric sequence (one that is formed by multiplying the same constant repeatedly to an initial value) $a_n = a_1 r^{(n-1)}$ where a_1 is the first term, r is the common ratio, and n is the number of terms in the sequence.

2. Convergent sequences—a sequence converges if it approaches a number. A sequence can be thought of as a function whose domain is the set of positive integers. As such, the concept of limit of a sequence is the same as the concept of limit of a function.

   For example, $\lim\limits_{n \to \infty} \left\{ \dfrac{n}{2n+3} \right\} = \dfrac{1}{2}$.

3. Divergent Sequences—a sequence is divergent if it does not approach a particular number; that is, it approaches $\pm\infty$.

   For example, $\lim\limits_{n \to \infty} \left\{ \dfrac{n^2}{2n+3} \right\} = \infty$.

II. **SERIES**—a series is the sum of the terms of a sequence.

A series converges if the sequence of its partial sums converges. For $\sum\limits_{k=1}^{\infty} a_k = a_1 + a_2 + a_3 + a_4 + \cdots + a_k + \ldots$, the sequence of partial sums is given by $\{S_k\}_1^\infty$ where $S_1 = a_1$, $S_2 = a_1 + a_2$, $S_3 = a_1 + a_2 + a_3$, ... , $S_k = a_1 + a_2 + a_3 + \cdots + a_k$. With most series, it is possible only to figure out whether it converges (or diverges) but not to figure out the actual sum. In general, the series for which it is possible to find the sum, if it exists, are geometric series and telescoping series.

A. Types of infinite series

1. *Geometric series*—this series is of the form $a + ar + ar^2 + \cdots + ar^n + \ldots \sum\limits_{k=0}^{\infty} ar^k$. This series converges (that is, its sum exists) if and only if $|r| < 1$ (that is, $-1 < r < 1$). If it converges, its sum is given by $S_\infty = \dfrac{a}{1-r}$.

2. *p-series*, $\sum\limits_{k=1}^{\infty} \dfrac{1}{k^p}$, $p > 0$, converges when $p > 1$ and diverges when $0 < p \leq 1$.

3. *Alternating series* are series with terms whose signs alternate. They are of the form $\sum\limits_{k=1}^{\infty} (-1)^k a_k$ or $\sum\limits_{k=1}^{\infty} (-1)^{k+1} a_k$.

4. *Harmonic series*, $\sum\limits_{k=1}^{\infty} \dfrac{1}{k}$, diverges. This is a *p*-series with $p = 1$.

5. *Alternating Harmonic series*, $\sum\limits_{k=1}^{\infty} \dfrac{(-1)^k}{k}$, converges.

6. *Alternating p-series* $\sum\limits_{k=1}^{\infty} \dfrac{(-1)^k}{k^p}$ converges for $p > 0$.

7. *Power series* in x, $\sum\limits_{k=1}^{\infty} a_k x^k$. Power series in $(x - a)$, $\sum\limits_{k=1}^{\infty} a_k (x - a)^k$. (More on power series later on.)

8. *Telescoping series* is a series in which all but a finite number of terms cancel out. It is either decomposed into partial fractions or you need to decompose it yourself. For example,

$$\sum_{k=1}^{\infty} \frac{1}{k^2 + 3k + 2} = \sum_{k=1}^{\infty} \left(\frac{1}{k+1} - \frac{1}{k+2} \right) = \left(1 - \frac{1}{2} \right) + \left(\frac{1}{2} - \frac{1}{3} \right) +$$

$$\left(\frac{1}{3} - \frac{1}{4} \right) + \cdots + \left(\frac{1}{k} - \frac{1}{k+1} \right) + \left(\frac{1}{k+1} - \frac{1}{k+2} \right) = 1 - \frac{1}{k+2}. \text{ This}$$

series converges because $\lim\limits_{k \to \infty} \left(1 - \dfrac{1}{k+2} \right) = 1$.

Test Tip

Just because a fraction can be decomposed into its partial fractions, does not mean it will be telescoping! Not all telescoping series converge.

For example: $\sum\limits_{k=1}^{\infty} k-(k+1) = (1-2) + (2-3) + (3-4) + \cdots +$

$k - (k+1) = -k$ and $\lim\limits_{k\to\infty} -k = -\infty$, *so the series diverges.*

B. Convergence/Divergence Tests for Series

Let $\sum\limits_{k=1}^{\infty} a_k$ be an infinite series of positive terms. The series $\sum\limits_{k=1}^{\infty} a_k$ converges if and only if the sequence of partial sums, $\{S_k\}_1^{\infty}$, converges. Also, $\sum\limits_{k=1}^{\infty} a_k = \lim\limits_{k\to\infty} S_k$. That is, the sum of the series equals the limit of the sequence of partial sums. Also, if a series converges absolutely, then it converges. This means that if $\sum\limits_{k=1}^{\infty} |a_k|$ converges, then $\sum\limits_{k=0}^{\infty} a_k$ converges. For example, $\sum\limits_{k=1}^{\infty} \frac{(-1)^k}{k^2}$ converges because $\sum\limits_{k=1}^{\infty} \left|\frac{(-1)^k}{k^2}\right|$, or, equivalently, $\sum\limits_{k=1}^{\infty} \frac{1}{k^2}$ converges (p-series with $p > 1$).

1. Divergence Test

If $\lim\limits_{k\to\infty} a_k \neq 0$, the series $\sum\limits_{k=1}^{\infty} a_k$ diverges. The contrapositive of this statement, which is logically equivalent to the statement, is also very useful. That is, if $\sum\limits_{k=1}^{\infty} a_k$ converges, then $\lim\limits_{k\to\infty} a_k = 0$. In other words, if a series is convergent, its terms must approach zero. However, $\lim\limits_{k\to\infty} a_k = 0$ does not imply convergence.

i. The series $\sum\limits_{k=1}^{\infty} \frac{k}{\sqrt{k^2+1}}$ is divergent since $\lim\limits_{k\to\infty} \frac{k}{\sqrt{k^2+1}} =$ $\lim\limits_{k\to\infty} \frac{1}{\sqrt{1 + \frac{1}{k^2}}} = 1 \neq 0$.

ii. An example of a series in which the terms approach zero but which is not convergent is the harmonic series, $\sum\limits_{k=1}^{\infty}\dfrac{1}{k}$.

2. Ratio Test

(a) If $\lim\limits_{k\to\infty}\dfrac{a_{k+1}}{a_k} < 1$ then the series $\sum\limits_{k=1}^{\infty}a_k$ converges; (b) if $\lim\limits_{k\to\infty}\dfrac{a_{k+1}}{a_k} > 1$ the series diverges. If $\lim\limits_{k\to\infty}\dfrac{a_{k+1}}{a_k} = 1$, this test is inconclusive; use a different convergence test. Specifically, the Ratio Test does not work for p-series because in that case, $\lim\limits_{k\to\infty}\dfrac{a_{k+1}}{a_k} = 1$. Use this test mainly when a_k involves factorials or kth powers.

For example, the series $\sum\limits_{k=1}^{\infty}\dfrac{1}{k!}$ converges since

$$\lim\limits_{k\to\infty}\frac{\dfrac{1}{(k+1)!}}{\dfrac{1}{k!}} = \lim\limits_{k\to\infty}\frac{k!}{(k+1)!} = \lim\limits_{k\to\infty}\frac{k!}{(k+1)k!} = \lim\limits_{k\to\infty}\frac{1}{k+1} = 0 < 1.$$

Also, $\sum\limits_{k=1}^{\infty}\dfrac{k^k}{k!}$ diverges because $\lim\limits_{k\to\infty}\dfrac{\dfrac{(k+1)^{k+1}}{(k+1)!}}{\dfrac{k^k}{k!}} =$

$$\lim\limits_{k\to\infty}\frac{(k+1)^{k+1}}{(k+1)!}\cdot\frac{k!}{k^k} = \lim\limits_{k\to\infty}\frac{(k+1)^k}{k^k} = \lim\limits_{k\to\infty}\left(\frac{k+1}{k}\right)^k =$$

$$\lim\limits_{k\to\infty}\left(1+\frac{1}{k}\right)^k = e > 1.$$

3. Ratio test for absolute convergence

(a) If $\lim\limits_{k\to\infty}\left|\dfrac{a_{k+1}}{a_k}\right| < 1$ then the series $\sum\limits_{k=1}^{\infty}a_k$ converges; (b) if $\lim\limits_{k\to\infty}\left|\dfrac{a_{k+1}}{a_k}\right| > 1$ the series *diverges*. If $\lim\limits_{k\to\infty}\left|\dfrac{a_{k+1}}{a_k}\right| = 1$, this test is inconclusive, use a different convergence test. The series need not have positive terms and need not be alternating to use this test. If a series converges but not absolutely, it is said to converge conditionally. Notice that all that the absolute value sign does is make the negative disappear.

i. The series $\displaystyle\sum_{k=1}^{\infty}\frac{(-1)^k 2^k}{k!}$ converges absolutely since

$$\lim_{k\to\infty}\frac{2^{k+1}}{(k+1)!}\cdot\frac{k!}{2^k}=\lim_{k\to\infty}\frac{2}{(k+1)}=0<1.$$

ii. The series $\displaystyle\sum_{k=1}^{\infty}\frac{(-1)^k}{k}$ does not converge absolutely (that is,

$\displaystyle\sum_{k=1}^{\infty}\left|\frac{(-1)^k}{k}\right|$, which is equivalent to $\displaystyle\sum_{k=1}^{\infty}\frac{1}{k}$, does not converge

because it is the harmonic series) but it converges without the absolute values by the alternating series test (the terms, in absolute value, decrease and approach zero). Thus,

$\displaystyle\sum_{k=1}^{\infty}\frac{(-1)^k}{k}$ converges conditionally.

4. **Comparison Test**

Suppose $\displaystyle\sum_{k=1}^{\infty}a_k$ and $\displaystyle\sum_{k=1}^{\infty}b_k$ are series with positive terms.

(a) If $\displaystyle\sum_{k=1}^{\infty}b_k$ is convergent and $a_k\le b_k$ for all k, then $\displaystyle\sum_{k=1}^{\infty}a_k$ converges. (b) If $\displaystyle\sum_{k=1}^{\infty}b_k$ is divergent and $a_k\ge b_k$ for all k, then $\displaystyle\sum_{k=1}^{\infty}a_k$ diverges. Part (a) says that if the series with larger terms converges, then the series with smaller terms converges. Part (b) says that if the series with smaller terms diverges, then the series with larger terms diverges. This test only applies to series with non-negative terms. Use this as a last resort, as other tests are often easier to apply.

i. For example, to see if $\displaystyle\sum_{k=2}^{\infty}\frac{3k}{k^2-2}$ converges, we compare it to a similar (and, in this case, smaller) series, $\displaystyle\sum_{k=2}^{\infty}\frac{3k}{k^2}$. This

series diverges because it is equivalent to $3\displaystyle\sum_{k=2}^{\infty}\frac{1}{k}$, which is a

divergent harmonic series. Since the smaller series diverges, the larger (original) series diverges.

ii. To see if $\displaystyle\sum_{k=1}^{\infty} \frac{5k}{2k^3 + k^3 + 1}$ converges, we compare it to a similar (and, in this case, larger) series, $\displaystyle\sum_{k=1}^{\infty} \frac{5k}{2k^3}$. This series converges because it is equivalent to a convergent p-series, $\dfrac{5}{2}\displaystyle\sum_{k=1}^{\infty} \frac{k}{k^2}$. Since the larger series converges, the smaller (original) series converges.

5. **Limit Comparison Test**

Suppose $\displaystyle\sum_{k=1}^{\infty} a_k$ and $\displaystyle\sum_{k=1}^{\infty} b_k$ are series with positive terms. If $\displaystyle\lim_{k\to\infty} \frac{a_k}{b_k} = c$ where $0 < c < \infty$, then either both series converge or both series diverge.

i. To see if $\displaystyle\sum_{k=1}^{\infty} \frac{\sqrt{k}}{k^2 + k + 3}$ converges, compare it to

$\displaystyle\sum_{k=1}^{\infty} \frac{\sqrt{k}}{k^2} = \sum_{k=1}^{\infty} \frac{1}{k^{3/2}}$, which is a convergent p-series.

$\displaystyle\lim_{k\to\infty} \frac{\frac{\sqrt{k}}{k^2 + k + 3}}{\frac{\sqrt{k}}{k^2}} = \lim_{k\to\infty} \frac{k^2}{k^2 + k + 3} = 1$. Since $0 < 1 < \infty$ and the second series converges, then the first series also converges.

ii. To see if $\displaystyle\sum_{k=1}^{\infty} \frac{\pi^k + \sqrt{k}}{3^k + k^2}$ converges, compare it to $\displaystyle\sum_{k=1}^{\infty} \frac{\pi^k}{3^k} = \displaystyle\sum_{k=1}^{\infty} \left(\frac{\pi}{3}\right)^k$, which is a divergent $(r > 1)$ geometric series.

$\displaystyle\lim_{k\to\infty} \frac{\pi^k + \sqrt{k}}{3^k + k^2} \cdot \frac{3^k}{\pi^k} = 1$. Since $0 < 1 < \infty$ and the second series diverges, then the first series also diverges.

6. **Alternating Series Test**

If the alternating series $\displaystyle\sum_{k=1}^{\infty} (-1)^k a_k = -a_1 + a_2 - a_3 + a_4 - a_5 + a_6 - \ldots$ where $a_k > 0$ for all k, satisfies (a) $a_k > a_{k+1}$ <u>and</u>

(b) $\lim\limits_{k\to\infty} a_k = 0$, then the series converges. If one of these conditions is not satisfied, the series diverges. That is, if each term is smaller than the previous term (in absolute value), and the terms are approaching zero, then the series converges. This applies only to alternating series. <u>Remainder:</u> $|R_k| \le a_{k+1}$. That is, when adding the first n terms of an alternating series, the remainder (or the error) is less than or equal to the first omitted term.

$\sum\limits_{k=1}^{\infty} \dfrac{(-1)^k}{k}$ converges because, a) $\dfrac{1}{k} > \dfrac{1}{k+1}$ for all k, and

b) $\lim\limits_{k\to\infty} \dfrac{1}{k} = 0$.

i.   Find the number of terms required to approximate the sum of the series $\sum\limits_{n=1}^{\infty} (\cos \pi n)\dfrac{1}{n!}$ with an error of less than 0.001.

$\sum\limits_{n=1}^{\infty} (\cos \pi n)\dfrac{1}{n!} = -\dfrac{1}{1} + \dfrac{1}{2} - \dfrac{1}{6} + \dots$ This is an alternating series whose terms are getting smaller. The first factorial greater than 1,000 is 7!, so $\dfrac{1}{7!} < 0.001$. Therefore, it takes 6 terms to calculate $\sum\limits_{n=1}^{\infty} (\cos \pi n)\dfrac{1}{n!}$ with an error of 0.001.

7. **Integral Test**
    Let $f(x)$ be a continuous, positive, decreasing function on $[1, \infty)$ which results when k is replaced by x in the formula for a_k. Then the series $\sum\limits_{k=1}^{\infty} a_k$ converges if and only if the improper integral $\int\limits_{1}^{\infty} f(x)dx$ converges.

    Use this test when $f(x)$ is easy to integrate. This test only applies to series with positive terms.

    $\sum\limits_{k=1}^{\infty} \dfrac{1}{k}$ diverges because $\int\limits_{1}^{\infty} \dfrac{1}{x}dx = \lim\limits_{l\to\infty}(\ln(x))\Big|_{1}^{l} = \infty$. That is, since the integral diverges, so does the series.

Keep in Mind...

➤ Do not confuse sequences with series—a series is the sum of the terms of a sequence.

➤ A series which converges absolutely, converges. This is confusing, but all it means is that if the series of absolute values of the terms converges, then the series itself converges. That is, if $\sum |u_k|$ converges, so does $\sum u_k$.

➤ Not every telescoping series converges!

CHAPTER 16
PRACTICE PROBLEMS

(See solutions on page 216)

1. $\lim\limits_{n \to \infty} \left\{ \dfrac{2n-1}{3en^2 - 1} \right\} =$

2. State whether or not the series converges and name the test used:

   (A) $\sum\limits_{k=1}^{\infty} \dfrac{\cos k}{k^2}$

   (B) $\sum\limits_{k=1}^{\infty} \dfrac{e^k}{k+1}$

   (C) $\sum\limits_{k=1}^{\infty} \dfrac{(-1)^k}{(k-1)!}$

Taylor and Maclaurin Series

I. POWER SERIES

1. For a power series $\sum_{k=0}^{\infty} c_k(x-a)^k$ exactly one of the following holds:

(a) The series converges only for $x = a$.

(b) The series converges absolutely (and hence converges) for all real values of x.

(c) The series converges absolutely (and hence converges) for all x in some finite open interval $(a - R, a + R)$ and diverges if $x < a - R$ or $x > a + R$.

At either of the points $x = a - R$ or $x = a + R$, the series may converge absolutely, converge conditionally, or diverge. The interval $(a - R, a + R)$ is called the interval of convergence and half of its size is called the radius of convergence of the series.

The radius of convergence is represented by $R = \lim_{n \to \infty} \left| \dfrac{c_{n+1}}{c_n} \right|$. The

center of the series is $x = a$. To find the interval of convergence, use the Ratio Test for Absolute Convergence.

i. For example, to find the interval of convergence of

$\sum_{k=1}^{\infty} \dfrac{(x-5)^k}{k^2}$, we apply the Ratio Test for Absolute

Convergence: $\lim_{k \to \infty} \left| \dfrac{a_{k+1}}{a_k} \right| = \lim_{k \to \infty} \left| \dfrac{(x-5)^{k+1}}{(k+1)^2} \cdot \dfrac{k^2}{(x-5)^k} \right| = \lim_{k \to \infty} |x - 5|.$

For the series to converge, $|x - 5| < 1$. That is, $-1 < x - 5 < 1$, which implies that $4 < x < 6$. Remember that the series might or might not be convergent at the endpoints, so this must be checked by substituting each endpoint into the original

series. For $x = 4$, the series becomes $\sum_{k=1}^{\infty} \frac{(-1)^k}{k^2}$ which

converges absolutely and therefore converges. We know this

because it is an alternating p-series with $p > 0$; also, $\sum_{k=1}^{\infty} \left| \frac{(-1)^k}{k^2} \right|$

is equivalent to the convergent p-series $(p > 1)$ $\sum_{k=1}^{\infty} \frac{1}{k^2}$.

For $x = 6$, the series becomes $\sum_{k=1}^{\infty} \frac{1}{k^2}$ which is a convergent

p-series $(p > 1)$. So, the interval of convergence for

$\sum_{k=1}^{\infty} \frac{(x-5)^k}{k^2}$ is [4, 6] and the radius of convergence is 1.

ii. To find the interval of convergence of $\sum_{k=0}^{\infty} k!x^k$ we

apply the Ratio Test for Absolute Convergence:

$\lim_{k \to \infty} \left| \frac{a_{k+1}}{a_k} \right| = \lim_{k \to \infty} \left| \frac{(k+1)!x^{k+1}}{k!x^k} \right| = \lim_{k \to \infty}(k+1)\,|\,x\,| = \infty$. This means that

the series diverges for all nonzero values of x. So it converges
only at $x = 0$, and the radius of convergence is $R = 0$.

II. **TAYLOR SERIES** can be used to approximate a function, $f(x)$,
by a polynomial of specified degree in the vicinity of a given point,
$x = a$. This is an extension of finding the equation of the tangent
line to a function at a point. Just like the tangent line approximates
the function fairly well in a small vicinity of the tangency point,
the Taylor polynomial approximates the function, $f(x)$, in the
vicinity of $x = a$. The difference between approximating the values
of a function using a tangent line at a point and using a Taylor
polynomial, is that the Taylor polynomial can be more accurate in a
larger vicinity of the point. The higher the degree of the polynomial,
the better the approximation.

1. If f has derivatives of all orders at $x = a$, then the Taylor series

for f about $x = a$ is given by: $f(n) = \sum_{n=0}^{\infty} \frac{f^{(n)}(a)}{n!}(x-a)^n = f(a) +$

$$f'(a)(x-a)+\frac{f''(a)}{2!}(x-a)^2+\frac{f'''(a)}{3!}(x-a)^3+\cdots+\frac{f^n(a)}{n!}(x-a)^n+\ldots$$

In the special case in which $a = 0$, the Taylor series is called the Maclaurin series for f. In that case, $\sum_{k=0}^{\infty}\frac{f^{(n)}(a)}{n!}(x)^n = f(0)+$

$$f'(0)(x)+\frac{f''(0)}{2!}(x)^2+\frac{f'''(0)}{3!}(x)^3+\cdots+\frac{f^n(0)}{n!}(x)^n+\ldots$$

i. For example, find the Taylor series of degree 3 for $f(x) = \frac{1}{x}$ about $x = 1$. First find the derivatives of f: $f'(x) = -\frac{1}{x^2}$, $f''(x) = \frac{2}{x^3}$, $f'''(x) = -\frac{6}{x^4}$. Next, evaluate the function and its derivatives at $x = 1$: $f(1) = 1$, $f'(1) = -1$, $f''(1) = 2 = 2!$, $f'''(1) = -6 = -3!$. Now substitute all the values into the Taylor series above to get: $P_3(x) = 1 - (x-1) + (x-1)^2 - (x-1)^3$. As seen in the graph below, the Taylor polynomial approximates $f(x)$ in the vicinity of $x = 1$.

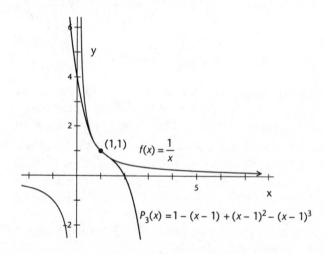

ii. Find the Maclaurin series of degree four for $f(x) = e^x$. A Maclaurin polynomial is a Taylor polynomial centered at $x = 0$. Since $f(x) = f'(x) = f''(x) = f'''(x) = f^{iv}(x)$, it follows that $f(0) = f'(0) = f''(0) = f'''(0) = f^{iv}(0) = 1$. Substituting into

the Maclaurin series, we have: $P_4(x) = 1 + x + \dfrac{x^2}{2!} + \dfrac{x^3}{3!} + \dfrac{x^4}{4!}$.

iii. Common Maclaurin series (and their intervals of convergence) that must be memorized:

$$\frac{1}{1-x} = 1 + x + x^2 + x^3 + \cdots = \sum_{k=0}^{\infty} x^k \quad \text{for } -1 < x < 1$$

$$e^x = 1 + x + \frac{x^2}{2!} + \frac{x^3}{3!} + \frac{x^4}{4!} + \cdots = \sum_{k=0}^{\infty} \frac{x^k}{k!} \quad \text{for } -\infty < x < \infty$$

$$\sin(x) = x - \frac{x^3}{3!} + \frac{x^5}{5!} - \frac{x^7}{7!} + \cdots = \sum_{k=0}^{\infty} (-1)^k \frac{x^{2k+1}}{(2k+1)!} \quad \text{for } -\infty < x < \infty$$

$$\cos(x) = 1 - \frac{x^2}{2!} + \frac{x^4}{4!} - \frac{x^6}{6!} + \cdots = \sum_{k=0}^{\infty} (-1)^k \frac{x^{2k}}{(2k)!} \quad \text{for } -\infty < x < \infty$$

$$\ln(1+x) = x - \frac{x^2}{2} + \frac{x^3}{3} - \frac{x^4}{4} + \cdots = \sum_{k=0}^{\infty} (-1)^k \frac{x^{k+1}}{(k+1)} \quad \text{for } -1 < x \leq 1$$

$$\tan^{-1}(x) = x - \frac{x^3}{3} + \frac{x^5}{5} - \frac{x^7}{7} + \cdots = \sum_{k=0}^{\infty} (-1)^k \frac{x^{2k+1}}{2k+1} \quad \text{for } -1 \leq x \leq 1.$$

2. Creating new power series from known power series. This is done by substituting x with a different quantity—in the known series expansion as well as in the interval of convergence!

i. For instance, a Maclaurin series for e^{-x^2} can be obtained by substituting $-x^2$ for x in the Maclaurin series for e^x. So,

$$e^{-x^2} = 1 + (-x^2) + \frac{(-x^2)^2}{2!} + \frac{(-x^2)^3}{3!} + \frac{(-x^2)^4}{4!} + \cdots = 1 - x^2 +$$

$\dfrac{x^4}{2!} - \dfrac{x^6}{3!} + \dfrac{x^8}{4!} + \ldots.$ Substituting x with $-x^2$ in the interval of

convergence for e^x, yields $-\infty < -x^2 < \infty \rightarrow 0 < x^2 < \infty \rightarrow$
$-\infty < x < \infty$.

ii. Also, we can write a Maclaurin series for $\dfrac{x}{1+x^3}$ by substituting

$-x^3$ for x in the Maclaurin series for $\dfrac{1}{1-x}$ and then multiplying the series by x: $\dfrac{1}{1+x^3} = 1 + (-x^3) + (-x^3)^2 + (-x^3)^3 + \cdots =$

$1 - x^3 + x^6 - x^9 + \cdots$. Multiplying by x, we get

$\dfrac{x}{1+x^3} = x - x^4 + x^7 - x^{10} + \cdots$. Substituting $-x^3$ into the interval

of convergence for $\dfrac{1}{1-x}$, yields $-1 < -x^3 < 1 \rightarrow -1 < x < 1$.

iii. The power series for $\tan^{-1}(2x)$ can be obtained by substituting

x with $2x$ in the power series for $\tan^{-1}(x)$: $\tan^{-1}(2x) = 2x - \dfrac{(2x)^3}{3} +$

$\dfrac{(2x)^5}{5} - \dfrac{(2x)^7}{7} + \cdots = 2x - \dfrac{8x^3}{3} + \dfrac{32x^5}{5} - \dfrac{128x^7}{7} + \cdots$. Substituting

x with $2x$ in the interval of convergence of $\tan^{-1}(x)$ yields:

$-1 \le 2x \le 1 \rightarrow -\dfrac{1}{2} \le x \le \dfrac{1}{2}$.

3. Differentiating and Integrating power series is done term by term. The interval and radius of convergence remain unchanged.

i. For instance, to show that $\dfrac{d}{dx}(\sin(x)) = \cos(x)$, we

differentiate the Maclaurin series for $\sin(x)$ term by term:

$\dfrac{d}{dx}(\sin(x)) = \dfrac{d}{dx}\left(x - \dfrac{x^3}{3!} + \dfrac{x^5}{5!} - \dfrac{x^7}{7!} + \cdots\right) = 1 - \dfrac{x^2}{2!} + \dfrac{x^4}{4!} - \dfrac{x^6}{6!} + \cdots =$

$\cos(x)$.

ii. To show that $\displaystyle\int \dfrac{1}{1+x^2}dx = \tan^{-1}(x) + C$, integrate the

Maclaurin series for $\dfrac{1}{1+x^2}$. We must first create this series by

substituting x with $-x^2$ in the Maclaurin series for $\dfrac{1}{1-x}$:

$\dfrac{1}{1+x^2} = 1 + (-x^2) + (-x^2)^2 + (-x^2)^3 + \cdots = 1 - x^2 + x^4 - x^6 + \cdots$.

Integrating this series term by term, yields: $x - \dfrac{x^3}{3} + \dfrac{x^5}{5} -$

$\dfrac{x^7}{7} + \cdots + C = \tan^{-1}(x) + C$ since $\tan^{-1}(0) = 0$, $C = 0$. So,

$\tan^{-1}(x) = x - \dfrac{x^3}{3} + \dfrac{x^5}{5} - \dfrac{x^7}{7} + \cdots$.

4. Lagrange's form of the remainder: $R_n(x) = \dfrac{f^{(n+1)}(c)(x-a)^{n+1}}{(n+1)!}$

where c is between a and x. This represents the error that arises when approximating a function with a Taylor polynomial of degree n. This is used to prove that a certain function is approximated by a series for all x-values.

i. To show that the Maclaurin series for $\cos(x)$ converges to $\cos(x)$ for all x, we must show that $\lim\limits_{x \to \infty} R_n(x) = 0$. Now, $f^{(n+1)}(x) = \pm\cos(x)$ or $f^{(n+1)}(x) = \pm\sin(x)$. In either case, for any value of c, $f^{(n+1)}(x) \le 1$. In this case, $a = 0$ so

$$|R_n(x)| = \left|\frac{f^{(n+1)}(x-a)^{n+1}}{(n+1)!}\right| \le \frac{|x|^{n+1}}{(n+1)!} \to 0 \text{ as } x \to \infty.$$

ii. Lagrange's form of the remainder, $R_n(x) = \dfrac{f^{(n+1)}(c)(x-a)^{n+1}}{(n+1)!}$, is also used when performing computations using Taylor series.

For instance, suppose we try to approximate $\sin\left(\dfrac{\pi}{3}\right)$ such that $R_n(x) \le 0.00005$. We'll use the Maclaurin series for $\sin(x)$, hence, $a = 0$. Since $f^{(n+1)}(x) = \pm\cos(x)$ or $f^{(n+1)}(x) = \pm\sin(x)$,

$f^{(n+1)}(x) \le 1$. Thus, $|R_n(x)| = \left|\dfrac{f^{(n+1)}(c)(x-a)^{n+1}}{(n+1)!}\right| \le \dfrac{|x|^{n+1}}{(n+1)!}$.

Since $x = \dfrac{\pi}{3}$, we must find an n value that satisfies

$$\left|R_n\left(\frac{\pi}{3}\right)\right| \le \frac{\left|\frac{\pi}{3}\right|^{n+1}}{(n+1)!} < 0.000005. \text{ By trial and error, } n = 8.$$

Therefore, $\sin\left(\dfrac{\pi}{3}\right) = \dfrac{\pi}{3} - \dfrac{\left(\frac{\pi}{3}\right)^3}{3!} + \dfrac{\left(\frac{\pi}{3}\right)^5}{5!} - \dfrac{\left(\frac{\pi}{3}\right)^7}{7!} = .8660212717.$

Using the calculator, $\sin\left(\dfrac{\pi}{3}\right) = .8660254038.$

Keep in Mind...

➤ Memorize the basic Maclaurin series because you will not have time to derive them on the AP exam.

➤ Apply the ratio test for absolute convergence to a series when finding the interval or radius of convergence.

➤ Don't forget to test the endpoints of the interval of convergence to see if the series is convergent or not at these points.

CHAPTER 17
PRACTICE PROBLEMS

(See solutions on page 217)

1. What is the difference between a Taylor polynomial and a Maclaurin polynomial?

2. Find the interval and radius of convergence of $\sum_{k=1}^{\infty} \dfrac{(2-x)^k}{k^3}$.

3. Find a Taylor series of degree four for $y = 3e^x$ at $x = 1$.

4. Find a Maclaurin series of degree six for $y = \dfrac{\sin(2x)}{x}$.

PART VI

THE EXAM

The Graphing Calculator

This chapter will describe how to use the Texas Instruments TI-83 Plus graphing calculators to solve calculus problems. The TI-83 Plus is very similar to the TI-83 and TI-84 so the steps that follow can be used for them as well.

The standard window is [–10, 10] for both axes. If this is not already set on your calculator, you can set it manually by clicking (WINDOW) and changing X_{min} and Y_{min} to –10 and changing X_{max} and Y_{max} to 10. You can also set the standard window by clicking (ZOOM) (6). The window you need to use can be dictated by the problem—if you are told the domain of x, then use that; or if you are graphing a trigonometric function, you must use the trig window, (ZOOM) (7). Generally, it will be up to you to figure out which window is best for a particular problem. With practice, you will become an expert at this.

 I. **EVALUATING FUNCTIONS, FINDING ZEROS, MAX/MIN POINTS AND INTERSECTION POINTS**

A. Evaluating a function at a point—that is, finding the y-value given an x-value. For instance, evaluate $f(3)$ given that $f(x) = 3e^x - x^2$.

1. Method 1. By brute force, you can replace x with 3:

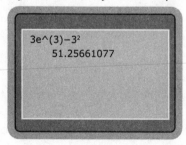

```
3e^(3)-3²
        51.25661077
```

2. Method 2. You may store 3 into x and then type in $3e^x - x^2$. To do this, press ③ (STO →) (X) (ENTER) (2nd) (LN) (X) (⟩) (-) (LN) (X) (x²) (ENTER).

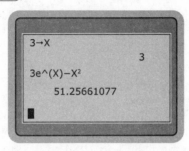

3. Method 3. Another way to substitute x with 3 is to use the CALCULATE menu to get:

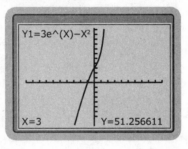

To do this, first enter the function in (Y=) then press (2nd) (Trace) ① ③ (ENTER).

Notice that the y-value in this case is rounded off to 6 decimal places instead of 8. However, this will not matter on the exam since you are required to use at least three-decimal-place accuracy. Also, when using this method, make sure that the number you are substituting for x is on the x-axis; otherwise you will get an error message that looks like this:

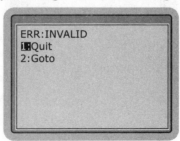

For instance, if you wanted to evaluate $f(13)$, but your window only contains x values from -10 to 10, you will get the above error message.

4. Method 4. You can evaluate $f(3)$ by looking at the table of values. To do this, first enter the function in (Y₁), then press (2nd) (GRAPH) and scroll to $x = 3$.

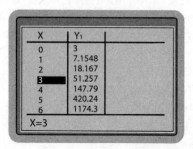

Notice that the y-value is rounded off to 3 decimal places—this is okay if this number is your final answer, but not if you need it in an earlier step. Remember, you are to round off—if you'd like to or if the problem asks you to—only at the very end of your calculations, otherwise you will accumulate round-off errors. You can also set the table of values to allow you to enter the x-value you want without scrolling. To do this, press (2nd) (WINDOW). Highlight *Ask* in the first line. After this, when you look at the table, it will be empty. This is because it is waiting for you to enter x-values.

Method 5. You can have the calculator substitute x with 3 for you:

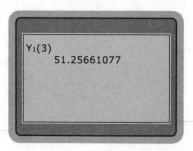

To do this, first enter the function in (Y=), then press (VARS) (Y-VARS) (1) (1) (((3) () (ENTER). (These steps assume that the function was entered in Y₁.) If the function was entered in Y₂ then the directions would be (VARS) (Y-VARS) (1) (2) (((3) () (ENTER).

B. Finding the zeros (also known as roots or *x*-intercepts) of a function. This feature will be used mostly to find critical points (roots of a derivative function) or inflection points (roots of a second derivative function). To illustrate, find the roots of $f(x) = 3e^x - x^2$.

1. Enter the function in (Y=) then press (2nd) (TRACE) (2). The calculator will ask you to enter a left bound. This is an *x* value less than the root. In this case, it looks like the root is somewhere between 0 and –2. So press (–2) and (ENTER) for the left bound.

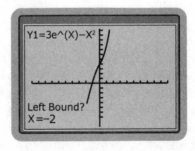

When you are asked for a right bound, that is, a number greater than the root, press (0) and (ENTER).

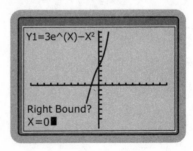

You are next asked to guess. Disregard this, just press (ENTER).

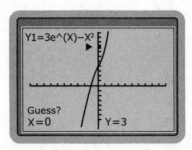

The answer is below:

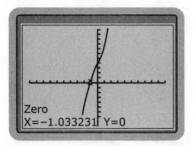

If you need to find more than one root, you must repeat this process for each one; you cannot find all roots at the same time. DO NOT use the TRACE button to approximate roots. It almost never works!

C. Finding the maximum/minimum points of a function.

1. To find the relative maximum point of $f(x) = 2x^4 + 7x^3 - 5x^2 - 10x + 3$, enter it in (Y=), then press (2nd) (TRACE) (4). For the left bound press (-2) and (ENTER).

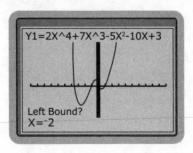

When asked for the right bound, press ⓪ and (ENTER).

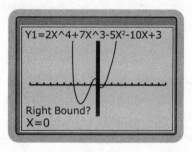

When asked to guess, disregard and press (ENTER). And the answer is:

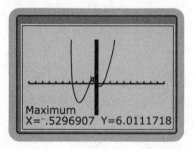

Note that (−.5296907, 6.0111718) is a relative maximum of *f*(*x*).

To find either minimum point, follow the steps above except, pick choice 3 instead of 4 from the CALCULATE menu. The absolute minimum is:

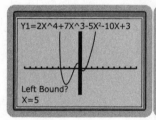 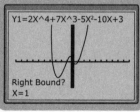 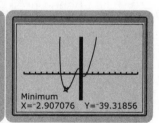

Note that when the answer has been found, the word "Minimum" appears. If you do not see this—minimum, maximum, zero, or intersection—then you have not found what you are looking for!

D. Finding the intersection points of two functions. This is used most often when finding the area between two curves or volumes of solids of revolution.

1. Find the intersection points of $f(x) = x^2$ and $g(x) = \sin(x)$. Enter the functions in (Y=) ($Y_1 = x^2$ and $Y_2 = \sin(x)$). Change the window to trig window by pressing (ZOOM) (7) and make sure that the calculator is in radian mode (click (MODE) and highlight Radian—99.9% of the time, your calculator needs to be in radian mode for the AP exam). Press (2nd) (TRACE) (5). Disregard the question "First Curve?", place the cursor as close to the intersection you are looking for as possible, and press (ENTER). Disregard the question "Second Curve" and press (ENTER). Disregard "Guess?" and press (ENTER). The given functions intersect at (0, 0) and (.87672622, .76864886). For short, after pressing (2nd) (TRACE) (5) and placing the cursor on the intersection, press (ENTER) three times to get the answer. Make sure you see the word INTERSECTION above the answer; otherwise you are not done. The graphs below were zoomed in for clarity's sake.

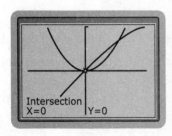
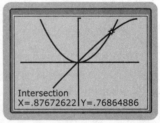

II. FINDING THE LIMIT OF A FUNCTION

A. To evaluate $\lim_{x \to a} f(x)$ you must replace x with values that are very close to a, from both sides of a. To evaluate $\lim_{x \to \pm\infty} f(x)$, you must replace x with very large values (if $x \to \infty$) or very low values (if $x \to -\infty$).

1. For instance, to evaluate $\lim\limits_{x \to 0} \frac{1}{x}$, enter $y = \frac{1}{x}$ into Y$_1$ and then evaluate it with values of x that approach zero from the left of zero:

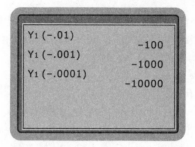

From here we see that $\lim\limits_{x \to 0^-} \frac{1}{x} = -\infty$. Checking the right-sided limit, we get:

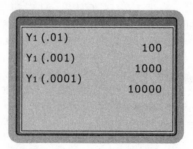

This shows that $\lim\limits_{x \to 0^+} \frac{1}{x} = +\infty$. Since the left and right-hand limits are not equal, we conclude that $\lim\limits_{x \to 0} \frac{1}{x}$ does not exist.

2. To evaluate $\lim\limits_{x \to \infty} \frac{1}{x}$, enter $y = \frac{1}{x}$ into Y$_1$, substitute x with large values:

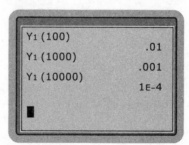

These calculations show that $\lim\limits_{x \to \infty} \dfrac{1}{x} = 0$. Note that you can press $\boxed{\text{2nd}}$ $\boxed{\text{ENTER}}$ after the first calculation to save time. This will copy your last step so all you need is to replace the 100 with 1000.

3. The limit of a function can also be evaluated by looking at its graph. From this graph we can see that $x = 0$ is a vertical asymptote of $y = \dfrac{1}{x}$ so $\lim\limits_{x \to 0} \dfrac{1}{x} = $ dne.

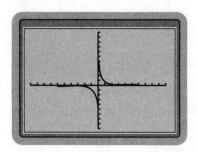

III. EVALUATING DERIVATIVES AND DRAWING THEIR GRAPHS

A. There are at least four ways to evaluate the derivative of a function at a given x-value. For instance, find $f'(-2)$ if $f(x) = \ln(1 - x)$.

1. You could find the derivative on your own and enter it in Y_1. That is, since $f'(x) = -\dfrac{1}{1-x}$ set $Y_1 = -\dfrac{1}{1-x}$ and, on the home screen, press $\boxed{\text{VARS}}$ $\boxed{\text{YVARS}}$ $\boxed{1}$ $\boxed{1}$ $\boxed{(\,}$ $\boxed{(\,}$ $\boxed{-2}$ $\boxed{)}$ $\boxed{\text{ENTER}}$.

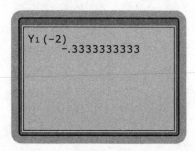

2. Or, you can let the calculator find the derivative and substitute in $x = -2$. In (Y=) enter $Y_1 = \ln(1-x)$. On the home screen (press (2nd) (Mode) to quit any screen you're in and go to the home screen), press (MATH) (8) (VARS) (YVARS) (1) (1) (,) (X) (,) (-2) ()) (ENTER). (Note that you must press the comma button after Y_1 and after X.)

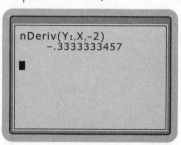

Note that this answer is not equivalent to the actual answer, $-\dfrac{1}{3}$, but it is a good enough approximation for the exam.

3. You can also enter (MATH) (8) (LN) (1) (−) (X) ()) (,) (X) (,) (-2) ()) (ENTER) to get the answer.

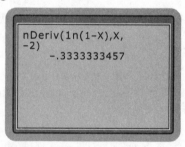

4. Once again, after graphing the function, you can use the CALCULATE menu. Press (2nd) (TRACE) (6) (-2) (ENTER).

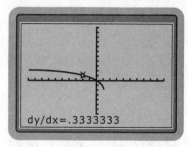

Again, make sure that the *x*-value you are entering is in your window, or else you will get an error message. In this case, any *x* value less than −10 or greater than 10 would produce an error.

B. Drawing the graph of $f'(x)$ and $f''(x)$.

1. To draw the graph of $f'(x)$, press (Y=) and enter $f(x)$ in Y_1. Then in Y_2 enter (MATH) (8) (VARS) (YVARS) (1) (1) (,) (X) (,) (X) ()) (ENTER). For instance, graph the derivative of $f(x) = x^4$.

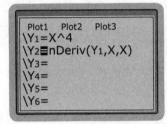
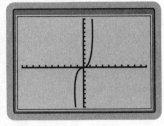

Notice that if you want to draw only the derivative graph, Y_2, you must disable the function graph, Y_1. (To disable an equation, place the cursor on the equal sign and press (ENTER). The equation will remain, but the calculator will not graph it. To enable the function, place the cursor on the equal sign and press (ENTER).) You can have both equations graphed at once, or even better, highlight one of them to tell the difference more easily:

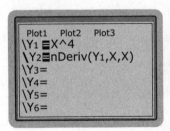
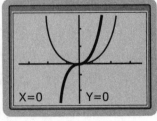

To highlight a function, place the cursor to the left of Y_2 and press (ENTER). To change it back, press (ENTER) 6 times. In this case, the derivative was highlighted. When graphing the above functions, the zoom-in feature was used for clarity.

2. To draw the graph of $f''(x)$ (given that the functions are entered as above in Y_1 and Y_2) enter the following in Y_3: (MATH) (8) (VARS) (YVARS) (1) (2) (,) (X) (,) (X) (⟩) (ENTER)). The graph of the second derivative is dotted.

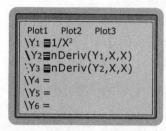
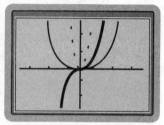

BIG DEAL: The calculator is by no means a perfect tool. Here are some instances to watch out for:

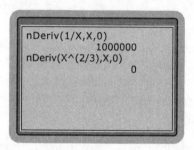

In the first instance, the calculator states that $f'(0) = 1000000$ for $f(x) = \dfrac{1}{x}$. We know this to be wrong since $f'(x) = -\dfrac{1}{x^2}$ and hence, $f'(0)$ does not exist.

In the second instance, the calculator states that $f'(0) = 0$ for $f(x) = x^{\frac{2}{3}}$. We know this not to be true because $f'(x) = \dfrac{2}{3x^{\frac{1}{3}}}$, and hence $f'(0)$ does not exist.

IV. EVALUATING DEFINITE INTEGRALS

A. There are three ways of evaluating definite integrals, which mostly occur when finding area or volume:

1. To evaluate $\int_{-2}^{3} x^2 dx$, on the home screen, enter (MATH) (9) (X) (x²) (,) (X) (,) (-2) (,) (3) (∫) (ENTER). Note that you must press the comma button after X², X, and –2.

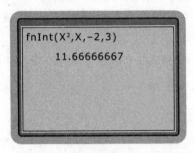

2. To evaluate $\int_{-2}^{3} x^2 dx$, enter x^2 in (Y=), more specifically in Y_1. Then, on the home screen, enter (MATH) (9) (VARS) (YVARS) (1) (1) (,) (X) (,) (-2) (,) (3) (∫) (ENTER).

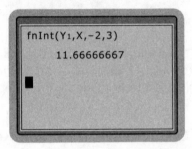

3. The third, more visual way to evaluate $\int_{-2}^{3} x^2 dx$, is to use the CALCULATE menu after having entered the function in (Y=). Press (2ⁿᵈ) (TRACE) (7) (-2) (3) (ENTER).

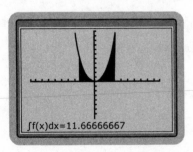

4. To evaluate an improper integral of the form $\int_{2}^{\infty}\frac{1}{x^2}dx$, let the upper limit get larger and larger by using the table of values. In this case, the last x in the expression entered in Y_2 stands for the upper limit of the integral. If we allow this upper limit to get larger and larger, 10, 100, 1000, and so on, the integral's value approaches 0.5. Evaluate $\int_{2}^{\infty}\frac{1}{x^2}dx$ by hand to verify that $\int_{2}^{\infty}\frac{1}{x^2}dx = 0.5$.

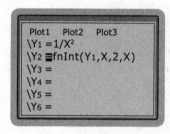
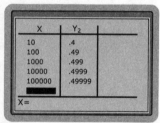

Note that the original function has been disabled so that the table only contains values of Y_2, those of the integral. Also, the table of values is in **Ask** mode for the independent variable.

 V. **USING NEWTON'S METHOD TO APPROXIMATE ROOTS OF FUNCTIONS**

A. Suppose we want to find the positive root of $f(x) = x^2 - 3$. Newton's Method uses the x-intercept of the tangent line at a given x-value to approximate the root of the function. The formula is: $x_{n+1} = x_n - \frac{f(x_n)}{f'(x_n)}$. Let $x_0 = 1$. Enter f in Y_1 and f' in Y_2 as shown below.

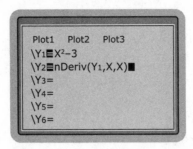

On the home screen, find the first approximation using the formula above and $x_0 = 1$.

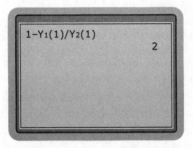

To find the next iteration, replace the 1 with ANS (in this case ANS = 2 but using ANS instead of 2 will make it much, much faster as you will see later on). To do this, press (2nd) (ENTER) and replace every 1 with ANS ((2nd) (() (-) ())).

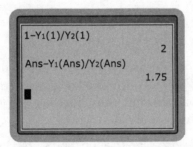

If you keep pressing (ENTER) you will get closer and closer to the actual root because the last statement gets calculated repeatedly and in every iteration **ANS** is substituted with the previous value.

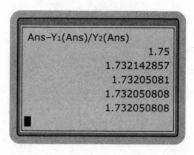

In this case, it took five iterations to get to 1.732050808, which is the actual root correct to 9 decimal places.

CALCULATOR TIPS:

To stop the calculator in the middle of graphing a function or in the middle of a calculation that takes too long, press (ON).

To go back to the previous line on the home screen, press (2nd) (ENTER). The calculator can memorize about 25 steps! So if you need to evaluate a function at more than one x-value, instead of re-entering the function every time, just press (2nd) (ENTER) and then substitute the new x-value.

Don't confuse the subtraction key with the negative key. They are not interchangeable!

The CLEAR button clears a whole statement; the DELETE button erases one character at a time.

To go to the beginning of a statement, press (2nd) (←).
To go to the end of a statement, press (2nd) (→).

To draw a vertical line, say $x = 6$, you need to use the Draw menu. Press (2nd) (PRGM) (4) (6). To erase any drawing, press (2nd) (PRGM) (1). You cannot evaluate, find roots, max/min points, intersection points, or derivatives of drawing objects.

To draw the tangent line to a function stored in Y_1 at a point, say $x = 3$, press (2nd) (PRGM) (5) (VARS) (Y-VARS) (1) (3) (ENTER). For instance, if you are asked to find the equation of the tangent line to a function at a point, you can graph the equation of the tangent line you've found and then draw it using the Draw menu to compare.

If you've made a mistake and need to erase a character and replace it with another, or if a character is missing, do not erase the whole statement; instead, use the insert feature, (2ⁿᵈ) (DEL). This creates an empty space for your new character.

To use the calculator with parametric equations, press (MODE) Par. Using the CALCULATE menu with parametric equations is very similar to using it with Cartesian equations. Make sure your t step is small, about 0.1, so the graph will not be jagged.

To use the calculator with polar equations, press (MODE) Pol. The CALCULATE menu is also simple to use. Just make sure that your θ_{step} is small, about 0.1.

ERR:SYNTAX means you typed in something wrong—perhaps an extra comma, a subtraction sign instead of a negative sign, etc. Generally, if you press 2, the calculator will place the cursor on the error and you can correct it.

ERR:INVALID DIM means that one of your plots in the (Y=) menu is active (it is highlighted). To deactivate it, place the cursor on it and press ENTER.

If you press (MODE) G-T, you can see the graph along with the table on the same screen! If you press (GRAPH), you can trace the curve and see the points in the table at the same time. If you press (2ⁿᵈ) (GRAPH), you can scroll down the x-values in the table.

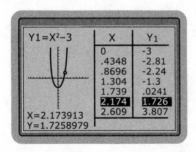

The Multiple-Choice Questions

Be sure to actively apply test-taking strategies when working on the multiple-choice section of the AP Calculus exams. These strategies include reading each question carefully, examining each answer choice, and carefully working through each exercise. By the way, the exact answer may not appear as a choice. If this happens, choose the answer that best approximates the exact answer.

The first section will include 45 multiple-choice questions to be completed in 105 minutes. The score on this section of the AP Calculus exams is based on the number of correctly answered questions. There will be no deductions for incorrect answers and no points will be given for unanswered questions. It is important, before taking any test, to keep in mind how it's scored. Since no points are deducted for incorrect answers, you should answer every question, even if you have to guess. Of course, do your best to eliminate some answer choices before guessing. Sometimes, the process of elimination leaves only one answer choice!

THE MULTIPLE-CHOICE SECTIONS

This multiple-choice portion of the test is divided into two parts:

Part A must be done in 60 minutes. This part has 30 questions. No calculator is allowed on this part.

Part B includes questions that may require the use of a graphing calculator. This part has 15 questions to be done in 45 minutes.

Example Question Found in Section 1, Part A:

The slope of the tangent line to the graph of $f(x) = 4x^3 - 3x^2 - x$ at $(-1, -6)$ is

    (A) -6

    (B) 5

    (C) 11

    (D) 17

Solution: The formula for the slope of the tangent line is the first derivative of the function. The first derivative of $f(x) = 4x^3 - 3x^2 - x$ is $f'(x) = 12x^2 - 6x - 1$. The slope of the tangent line of this function at the point $(-1, -6)$ is found by substituting -1 in for x in $f'(x)$. The result is $f'(-1) = 12(-1)^2 - 6(-1) - 1 = 12(1) - 6(-1) - 1 = 12 + 6 - 1 = 17$. The correct answer is D.

SECTION I, PART A: EXAMPLE II:

What is $\lim\limits_{h \to 0} \dfrac{\sin(\pi + h) - \sin(\pi)}{h}$?

    (A) -1

    (B) 0

    (C) 1

    (D) undefined

Solution: The given limit is actually the definition of the derivative. The function being used is $f(x) = \sin(x)$, while the specific value at which the derivative is being found is π. The derivative of the function is $f'(x) = \cos(x)$. Evaluating the derivative at π results in $f'(\pi) = \cos(\pi) = -1$. The correct answer is A.

Example Question Found in Section 1, Part B:

What is the inflection point of $f(x) = 4x^3 - 3x^2 - x$?

(A) $\left(\dfrac{-1}{4}, \dfrac{-3}{8}\right)$

(B) $\left(\dfrac{-1}{4}, \dfrac{3}{8}\right)$

(C) $\left(\dfrac{1}{4}, 0\right)$

(D) $\left(\dfrac{1}{4}, \dfrac{-3}{8}\right)$

Solution: Go ahead and use the graphing calculator, if you wish.

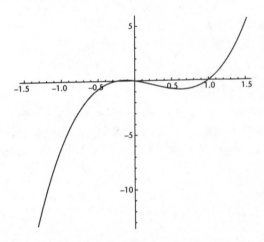

Looking at the graph of the function, it appears that an inflection point does exist because the concavity changes from concave down to concave up at $x = \dfrac{1}{4}$. So, you can eliminate choices A and B. It is evident from the graph that the y-value of the inflection point is negative. Substituting $x = \dfrac{1}{4}$ back into the original function, $f\left(\dfrac{1}{4}\right) = \dfrac{-3}{8}$.

Hence, the inflection point is $\left(\dfrac{1}{4}, \dfrac{-3}{8}\right)$. The correct answer is D.

SECTION I, PART B: EXAMPLE II:

$$\int \frac{6x - 2}{x^2 - 4x + 3}\,dx =$$

(A) $8 \ln|x - 3| + 2 \ln|x - 1| + C$

(B) $2 \ln\dfrac{|x - 3|^4}{|x - 1|} + C$

(C) $\dfrac{2}{3} \ln|x^2 - 4x + 3| + C$

(D) $\dfrac{3}{2} \ln|x^2 - 4x + 3| + C$

Solution: Attempting this as a substitution would lead nowhere. Because the denominator is factorable, decomposition of the fraction into partial fractions is in order. This is a Calculus BC problem.

$$\int \frac{6x - 2}{x^2 - 4x + 3}\,dx = \int \frac{6x - 2}{(x - 3)(x - 1)}\,dx = \int \left[\frac{A}{x - 3} + \frac{B}{x - 1}\right] dx$$

$$A = \frac{6(3) - 2}{3 - 1} = \frac{16}{2} = 8 \qquad B = \frac{6(1) - 2}{1 - 3} = \frac{4}{-2} = -2$$

$$\int \left[\frac{8}{x - 3} + \frac{2}{x - 1}\right] dx = 8 \ln|x - 3| - 2 \ln|x - 1|$$

This answer is correct, but does not match the choices; log rules can simplify it. Choice (B) now comes into view.

$$2(4\ln|x-3|-\ln|x-1|) = 2\ln\frac{|x-3|^4}{|x-1|}+C$$

The correct answer is (B).

Chapter 20

The Free-Response Questions

You need to work the free-response questions out, step by step. These questions often involve graphs. Partial credit is given for steps of these exercises. It is extremely important that you show all of the steps. Doing so allows the AP Reader to follow your logic and provides more opportunity for you to earn every point for which you are eligible. In fact, answers that appear without proper supporting logic usually earn no credit! Also, when explaining your process, use complete sentences.

Each part of every exercise has an indicated workspace. Show all work for that part in that space only! Be sure to write neatly, so the reader can follow your work! If you make an error, erase it or cross it out. Crossed-out work will not be scored.

The free-response section will include 6 problems to be completed in 90 minutes. You will use pencil or dark blue or black ink to write out your work on these questions. Although the parts of each question are not necessarily equally weighted, each question as a whole is equally weighted.

THE FREE-RESPONSE SECTIONS

The free-response section of the AP Calculus examinations has two parts. A graphing calculator is allowed on Part A, which has two questions. You will have 30 minutes to work on Part A.

Part B has 4 questions and use of the graphing calculator is prohibited. You have 60 minutes to work on this part. During this hour, you may continue to work on Part A, without the use of the graphing calculator.

EXAMPLE QUESTION FOUND IN SECTION II, PART A:

Let R be the region bounded by the graphs of $y = 10e^{-x}$, $y = \ln x$, $x = 1$, and $x = 2$.
   A. What is the area of the region bounded by the curves?
   B. Set up, but do not solve, the integral expression, in terms of a single variable, for the volume of the solid generated by revolving this enclosed region around the y-axis.
   C. Set up, but do not solve, the integral expression, in terms of a single variable, for the volume of the solid generated by revolving this enclosed region around the line $y = -1$.

Solution:

A. What is the area of the region bounded by the curves?

Take a look at the graph.

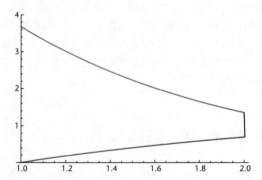

Knowing that the graph of $y = 10e^{-x}$ is above the graph of

$y = \ln x$, you set up your definite integral as $\int\limits_{1}^{2}(10e^{-x} - \ln x)dx$.

The smaller x-value is the lower limit of integration; the larger x-value is the upper limit of integration. The integrand is the top graph minus the bottom graph.

$$\int\limits_{1}^{2}(10e^{-x} - \ln x)dx \approx 1.939$$

This part would be worth 3 points: 1 for the graph, 1 for setting up the integral, and 1 for finding the numerical solution.

B. Set up, but do not solve, the integral expression, in terms of a single variable, for the volume of the solid generated by revolving this enclosed region around the y-axis.

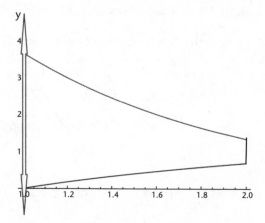

Finding the volume by revolving this region around the y-axis

would require the method of cylindrical shells: $V = 2\pi\int\limits_{a}^{b} rh\,dx$.

The radius is the distance between the y-axis and a cross-section: x. The height is the distance between the two functions: $10e^{-x} - \ln x$. The definite integral to evaluate this situation is:

$2\pi\int\limits_{1}^{2}(x)(10e^{-x} - \ln x)dx$. This part would be worth 1 point.

C. Set up, but do not solve, the integral expression, in terms of a single variable, for the volume of the solid generated by revolving this enclosed region around the line $y = -1$.

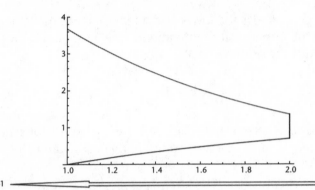

Finding the volume by revolving this region around the line $y = -1$ requires the method of washers: $V = \pi \int_a^b [R^2 - r^2]dx$.

The R in this exercise is the distance between $y = -1$ and the curve $y = 10e^{-x}$. So, $R = 1 + y = 1 + 10e^{-x}$. The r in this exercise is the distance between $y = -1$ and the curve $y = \ln x$, making $r = 1 + y = 1 + \ln x$. Putting this all together, you get:

$V = \pi \int_1^2 [(1+10e^{-x})^2 - (1+\ln x)^2]dx$. This part would be worth 1 point.

This previous example is worth 5 points total.

SECTION II, PART A: EXAMPLE II:

Let f be the function given by $f(x) = \frac{1}{2}x^3 - \frac{1}{2}x^2 - 1$.

A. Write an equation of the tangent line at $x = -1$.
B. List and identify all relative extreme points, both minimum and maximum.
C. What is the inflection point?

Solution:

A. Write an equation of the tangent line at $x = -1$.

The slope of the tangent line is the first derivative, $f'(x) = \frac{3}{2}x^2 - x$, evaluated at $x = -1$: $f'(-1) = \frac{3}{2}(-1)^2 - (-1) = \frac{3}{2} + 1 = \frac{5}{2}$. The point when $x = -1$ is $(-1, -2)$. Using the slope-intercept form of the equation of a line, $y = mx + b$, the b value is $\frac{1}{2}$. Hence, the equation of the tangent line is $y = \frac{5}{2}x + \frac{1}{2}$.

This part would be worth 2 points: 1 for the derivative and 1 for writing the equation of the tangent line.

B. List and identify all relative extreme, both minimum and maximum, points.

The first derivative set equal to 0 gives the critical numbers:

$$f'(x) = \frac{3}{2}x^2 - x = 0$$

$$x\left(\frac{3}{2}x - 1\right) = 0$$

$$x = 0 \quad \text{and} \quad x = \frac{2}{3}$$

The critical points are $(0, -1)$ and $\left(\frac{2}{3}, \frac{-29}{27}\right)$. The second derivative is $f'(x) = 3x - 1$. Evaluating the second derivative at $x = 0$, the result is negative, which indicates a relative maximum point. Evaluating the second derivative at $x = \frac{2}{3}$, the result is positive, which indicates a relative minimum point. This part would be worth 2 points: 1 for finding the critical points and 1 for identifying each.

C. What is the inflection point?

Setting the second derivative equal to 0, $f''(x) = 3x - 1 = 0$,

$x = \dfrac{1}{3}$. The inflection point is $\left(\dfrac{1}{3}, \dfrac{-28}{27}\right)$. This part would be

worth 1 point.

EXAMPLE QUESTION FOUND IN SECTION II, PART B:

The acceleration of a particle is $a(t) = -32$ ft/sec^2, the initial velocity of the particle is 64 ft/sec, and the initial height of the particle is 40 ft.
   A. What is the formula for the velocity of the particle at any time t?
   B. What is the formula for the position of the particle at any time t?
   C. What is the maximum height this particle reaches?

A. Given $a(t) = -32$, you should find its antiderivative as the velocity of the particle. The antiderivative of $a(t) = -32$ is $v(t) = -32t + C$. You know that the initial velocity is 64; that is $v(0) = 64$. This means

$$v(t) = -32t + C$$
$$v(0) = -32(0) + C = 64$$
$$C = 64$$
$$v(t) = -32t + 64$$

This should earn you 2 points, 1 for the antidifferentiation and 1 for finding the constant.

B. You now know that $v(t) = -32t + 64$. You take the antiderivative of $v(t)$ to get the position of the particle, $s(t)$, at any time t. The antiderivative is $s(t) = -16t^2 + 64t + C$. You know that the initial position, $s(0) = 40$. Hence,

$$s(t) = -16t^2 + 64t + C$$
$$s(0) = -16(0)^2 + 64(0) + C = 40$$
$$C = 40$$
$$s(t) = -16t^2 + 64t + 40$$

This exercise should also earn you 2 points, 1 for the anti-differentiation and 1 for finding this constant.

C. The maximum height of this particle is reached when its velocity is zero. (The particle instantaneously stops so that it can make its way back down.) Setting the velocity equal to zero, you get

$$v(t) = -32t + 64 = 0$$
$$-32t = -64$$
$$t = 2$$

You now know that the particle reaches its maximum height when $t = 2$. You need to find the maximum height. Place $t = 2$ into the position function:

$$s(t) = -16t^2 + 64t + 40$$
$$s(2) = -16(2)^2 + 64(2) + 40 = -16(4) + 64(2) + 40$$
$$= -64 + 128 + 40 = 104$$

You will earn 2 points for this exercise: 1 point for finding the time that the particle reaches its maximum height and 1 point for finding that maximum height.

The previous example is worth 6 points total.

SECTION II, PART B: EXAMPLE II:

Use $x^2 - xy + 4y^2 = 16$ for A – C.

    A. What is the slope in terms of x and y?

    B. What is the slope of this curve where $x = 0$ and $y = -2$?

C. What is $\dfrac{d^2y}{dx^2}$ at $x = 1$?

Solution:

A. What is the slope in terms of x and y?

$$x^2 - xy + 4y^2 = 16$$

$$2x - 1y + \frac{dy}{dx}(-x) + 8y\frac{dy}{dx} = 0$$

$$-x\frac{dy}{dx} + 8y\frac{dy}{dx} = -2x + y$$

$$\frac{dy}{dx}(-x + 8y) = -2x + y$$

$$\frac{dy}{dx} = \frac{-2x + y}{-x + 8y}$$

This would earn 2 points, 1 for the differentiation and 1 for the simplification.

B. What is the slope of this curve where $x = 0$ and $y = -2$?

The slope of the curve is the first derivative, $\dfrac{dy}{dx} = \dfrac{-2x + y}{-x + 8y}$.

At $(0, -2)$, the slope is $\dfrac{1}{8}$. This would earn 2 points, 1 for finding the y-coordinate of the point and 1 for finding the slope.

C. What is $\dfrac{d^2y}{dx^2}$ at $(0, -2)$?

Since $\dfrac{dy}{dx} = \dfrac{-2x + y}{-x + 8y}$, the second derivate is found by using the quotient rule:

$$\frac{d^2y}{dx^2} = \frac{(-x+8y)\left(-2+\dfrac{dy}{dx}\right)-(-2x+y)\left(-1+8\dfrac{dy}{dx}\right)}{(-x+8y)^2}$$

$$\frac{(-x+8y)\left(-2+\dfrac{-2x+y}{-x+8y}\right)-(-2x+y)\left(-1+8\left[\dfrac{-2x+y}{-x+8y}\right]\right)}{(-x+8y)^2}.$$

Evaluate this at $x = 0$, which is the point $(0, -2)$.

$$\frac{d^2y}{dx^2} = \frac{(-16)\left(-2+\dfrac{1}{8}\right)-(-2)\left(-1+8\left[\dfrac{1}{8}\right]\right)}{(-16)^2} = \frac{32-2+2(-1+1)}{(-16)^2}$$

$$= \frac{-32-2}{(-16)^2} = \frac{-34}{16(16)} = \frac{-17}{8(16)} = \frac{-17}{128}$$

This would earn 2 points, 1 for the differentiation and 1 for the evaluation.

AP CALCULUS SCORING

On the multiple-choice portion of the examination, total scores will be based upon the number of questions answered correctly. (There will be no deductions for incorrect answers and no points will be given for unanswered questions.) Before taking any examination, keep in mind how the test is scored. Since no points are deducted for incorrect answers, you should answer every question, even if you have to guess.

The scores for the multiple-choice and the free-response sections are equally weighted. Your score will be a combined score of the computer-scored multiple-choice portion and the AP Reader-scored free-response portion. The highest possible score on the actual

AP examination is 5. Each college or university determines the necessary score for credit in a college-level course.

As is common on standardized examinations, students are not expected to be able to answer every question.

If you take the AP Calculus BC examination, you will receive an AP Calculus AB examination subscore since approximately 60% of the Calculus BC examination is at the Calculus AB level.

CALCULATOR FOR AP CALCULUS

A calculator is allowed on certain sections of the AP Calculus exams. You should be prepared to use a graphing calculator that has the following capabilities: plotting the graph of a function, finding the zeros of functions, and numerically calculate derivatives and definite integrals. If you have questions about the appropriateness of your calculator for use on the AP Calculus examination you're taking, visit the AP Calculus AB & BC home page at *www.collegeboard.org*.

You must show the mathematical steps used to achieve the results, even if you use a calculator, as shown above with the area between the two curves. The graph is shown. The definite integral is shown, using standard calculus or mathematical notation. Also, the answer is clearly shown. If you use a calculator to determine the result, you need not "spell out" the steps followed if you were to do it by hand. In other words, simply showing the definite integral used to reach the result and showing the answer is adequate. Along these same lines, when you are asked to justify an answer, you need to mathematically justify; showing an outcome from a calculator is not sufficient!

In addition, when showing a graph, be sure to label it correctly and thoroughly.

What about approximating answers? You should have approximations correct to three decimal places, unless specified differently in an exercise. Do not round until the very end of the exercise!

SOLUTIONS FOR PRACTICE PROBLEMS

CHAPTER 2
SOLUTIONS

1. $y = \dfrac{1}{2}(x+1)^3 - 3$

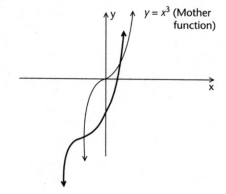

The mother function is widened, moved down 3 units and to the left one unit.

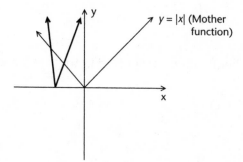

2. $y = 2|3x + 4|$

The mother function is narrowed and moved left $\dfrac{4}{3}$ units since $3x + 4 = 3\left(x + \dfrac{4}{3}\right).$

3. $y = \sqrt{x-6} + 1$

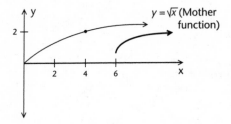

The mother function is moved 6 units right and one unit up.

4. $y = -\dfrac{3}{x} + 2$

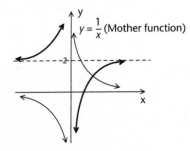

Mother function is reflected in the *x*-axis and moved up 2 units.

5. $y = e^{x+2} - 1$

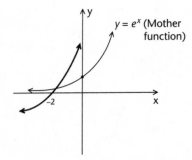

Mother function is moved 2 units left and one unit down.

6. $y = \ln(4 - x)$

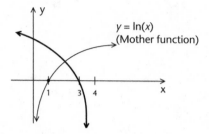

Mother function is reflected in y-axis and moved 4 units to the right.

CHAPTER 3
SOLUTIONS

1. (D) $-\dfrac{1}{8}$

$\lim\limits_{x \to \infty} \dfrac{1 - 3x + 6x^2 - x^{10}}{2 + 4x^4 - 8x^7 + 8x^{10}} = \lim\limits_{x \to \infty} -\dfrac{x^{10}}{8x^{10}} = -\dfrac{1}{8}$. Use ratio of leading terms.

2. (A) 0

$\lim\limits_{x \to \infty} \dfrac{\sin x}{x} = 0$

As $x \to \infty$, the denominator increases without bound whereas the numerator oscillates between -1 and 1.

3. (D) $-\infty$

$\lim\limits_{x \to 0^+} (\ln x) = -\infty$

Use the graph of $y = \ln x$.

4. (C) $\dfrac{1}{6}$

$$\lim_{x \to 3} \frac{\sqrt{x+6}-3}{x-3} = \frac{1}{6}$$

Recognize the limit as the derivative of $f(x) = \sqrt{x+6}$ at $x = 3$.

So $f'(x) = \dfrac{1}{2\sqrt{x+6}}$ and $f'(3) = \dfrac{1}{2\sqrt{9}} = \dfrac{1}{6}$.

5. (D) Does not exist

$$\lim_{x \to 1} \frac{3}{x-1} = dne$$

From the graph of $y = \dfrac{3}{x-1}$ we see that $\lim\limits_{x \to 1^-} \dfrac{3}{x-1} = -\infty$ and

$$\lim_{x \to 1^+} \frac{3}{x-1} = +\infty.$$

CHAPTER 4
SOLUTIONS

1. Vertical asymptote: $x = -3$; horizontal asymptote: $y = 3$

$$f(x) = \frac{3x^2 - 9x}{x^2 - 9} \quad f(x) = \frac{3x(x-3)}{(x+3)(x-3)} = \frac{3x}{x+3} \quad \text{vertical asymptote}$$

occurs when $x + 3 = 0$ or at $x = -3$.

$$\lim_{x \to \pm\infty} \frac{3x^2 - 9x}{x^2 - 9} = 3 \text{ horizontal asymptote: } y = 3.$$

2. Vertical asymptote: $x = \sqrt[3]{4}$; horizontal asymptote: $y = -1$

$$f(x) = \frac{x^3 + 3x^2 - 1}{4 - x^3} \text{ vertical asymptote occurs when } 4 - x^3 = 0 \text{ or at}$$

$x = \sqrt[3]{4}$ since $x = \sqrt[3]{4}$ is not a root of the numerator.

$\lim\limits_{x \to \pm\infty} f(x) = -1$ horizontal asymptote: $y = -1$.

3. (A) $y = 7$ is a horizontal asymptote of $f(x)$. The curve could have vertical asymptotes.

   (B) As $x \to -\infty$ $f(x)$ does not approach a horizontal asymptote, it decreases without bound. The curve could have vertical asymptotes.

   (C) $x = 4$ is a vertical asymptote of $f(x)$. The curve could have horizontal asymptotes.

CHAPTER 5
SOLUTIONS

1. $x < 1$

   $f(x) = \dfrac{2}{\sqrt{1-x}}$ is continuous when $1 - x > 0$ or $x < 1$.

2. Removable discontinuity at $x = -2$; nonremovable discontinuity at $x = 2$.

   $f(x) = \dfrac{x^2 + x + 6}{x^2 - 4} = \dfrac{(x+2)(x+3)}{(x+2)(x-2)}$ $\therefore$ $f(x)$ has a removable

   discontinuity at $x = -2$ and a nonremovable discontinuity at $x = 2$.

3. $x = -1, x = 0$

   Look at the graph of $f(x) = \begin{cases} 2 - x, & x < -1 \\ \dfrac{1}{x}, & -1 \le x \le 2 \\ \dfrac{1}{2}, & x > 2 \end{cases}$

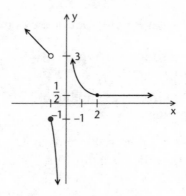

The function is discontinuous at $x = 0$ and $x = -1$.

CHAPTER 6
SOLUTIONS

1. $\begin{cases} x(t) = 2\sin t \\ y(t) = 3\cos t \end{cases} \rightarrow \begin{cases} \dfrac{x}{2} = \sin t \\ \dfrac{y}{3} = \cos t \end{cases} \rightarrow \begin{cases} \dfrac{x^2}{4} = \sin^2 t \\ \dfrac{y^2}{9} = \cos^2 t \end{cases} \rightarrow$

$\dfrac{x^2}{4} + \dfrac{y^2}{9} = 1.$ Ellipse $a = 2$, $b = 3$.

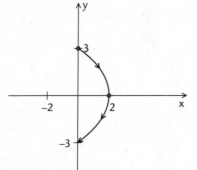

$0 \le t \le \pi$

t	0	$\pi/2$	π
x	0	2	0
y	3	0	-3

Don't forget to indicate direction of motion.

2. $r(t) = \left(\dfrac{t}{2}\right)i + (e^t)j \quad t > 0.$

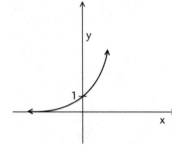

In parametric form, $\begin{cases} x(t) = \dfrac{t}{2} \\ y(t) = e^t \end{cases} \quad t > 0.$

In Cartesian form, $y = e^{2x}, \quad x > 0.$

3. $r = 2 - 3\cos\theta$ is a looped limaçon.

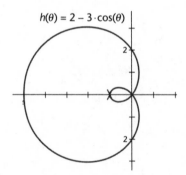

$h(\theta) = 2 - 3 \cdot \cos(\theta)$

4. $r = 4\sin(6\theta)$ has 12 petals, each with length 4.

5. $r = \cos\theta \rightarrow r^2 = r\cos\theta \rightarrow x^2 + y^2 = x \rightarrow x^2 - x + y^2 = 0 \rightarrow$

$x^2 - x + \dfrac{1}{4} + y^2 = \dfrac{1}{4} \rightarrow \left(x - \dfrac{1}{2}\right)^2 + y^2 = \dfrac{1}{4}.$ This is a circle with the

center at $\left(\dfrac{1}{2}, 0\right)$ and radius $\dfrac{1}{2}.$

6. $x = 2 \rightarrow r\cos\theta = 2 \rightarrow r = \dfrac{2}{\cos\theta} \rightarrow r = 2\sec\theta.$

CHAPTER 7
SOLUTIONS

1. $y' = \dfrac{23}{(1-5x)^2}$

$y' = \dfrac{(1-5x)(3) - (3x+4)(-5)}{(1-5x)^2} = \dfrac{3 - 15x + 15x + 20}{(1-5x)^2} = \dfrac{23}{(1-5x)^2}.$

2. $y' = \dfrac{5}{2\sqrt{5x+3}}$

$\dfrac{d}{dx}(\ln(e^{\sqrt{5x+3}})) = \dfrac{d}{dx}(\sqrt{5x+3}) = \dfrac{5}{2\sqrt{5x+3}}.$

3. y' at $x = -1$ is $\dfrac{1}{3}$

$3x - x^2y = 5y \rightarrow 3 - x^2y' - 2xy = 5y' \rightarrow 3 - 2xy = 5y' + x^2y' \rightarrow$

$3 - 2xy = y'(5 + x^2) \rightarrow y' = \dfrac{3 - 2xy}{5 + x^2}$. To evaluate y' at $x = -1$, we

must substitute $x = -1$ into the original equation to find y.

$3(-1) - (-1)^2 y = 5y \rightarrow -3 - y = 5y \rightarrow -3 = 6y \rightarrow y = -\dfrac{1}{2}.$

So $y'\Big|_{\left(-1, -\frac{1}{2}\right)} = \dfrac{3 - 2(-1)\left(-\dfrac{1}{2}\right)}{5 + (-1)^2} = \dfrac{2}{6} = \dfrac{1}{3}.$

4. Derivative does not exist.

$y = x^2 - 4x \xrightarrow{\text{INVERSE}} x = y^2 - 4y \xrightarrow{\text{DERIVATIVE}} 1 = 2yy' - 4y' \rightarrow 1 = y'(2y - 4)$. Since $x = 2$ is on the original function, $y = 2$ is on the inverse. So $y' = \dfrac{1}{2y - 4} \rightarrow y' = \dfrac{1}{2(2) - 4} = \dfrac{1}{0}$ dne. Graphically

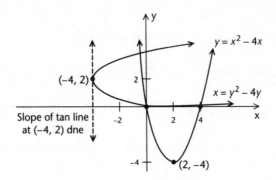

CHAPTER 8
SOLUTIONS

1. Critical points: $(0, 2)$, $\left(\pm\sqrt{\dfrac{3}{2}}, \dfrac{-1}{4} \right)$; inflection points: $\left(\pm\dfrac{\sqrt{2}}{2}, \dfrac{3}{4} \right)$.

   Relative maximum: $(0, 2)$, Absolute minimum y value: $y = -\dfrac{1}{4}$.

   Critical points:

   $$y' = 4x^3 - 6x = 0 \rightarrow y' = 2x(2x^2 - 3) = 0 \rightarrow x = 0, \pm\sqrt{\dfrac{3}{2}}.$$

   To find y-values, substitute x-values, into original function.
   $$y = (0)^4 - 3(0)^2 + 2 = 2 \; (0, 2)$$
   $$y = \left(\pm\sqrt{\dfrac{3}{2}} \right)^4 - 3\left(\pm\sqrt{\dfrac{3}{2}} \right)^2 + 2 = \dfrac{9}{4} - \dfrac{9}{2} + 2 = \dfrac{-1}{4} \quad \left(\pm\sqrt{\dfrac{3}{2}}, \dfrac{-1}{4} \right)$$

   Inflection points: $y'' = 12x^2 - 6 = 0 \rightarrow x = \pm\dfrac{\sqrt{2}}{2}.$

y''	+	0	–	0	+
x	-1	$-\dfrac{\sqrt{2}}{2}$	0	$\dfrac{\sqrt{2}}{2}$	1

Since there's a sign change at both $x = \dfrac{\sqrt{2}}{2}$ and $x = -\dfrac{\sqrt{2}}{2}$,

inflection points occur here. To find y values, substitute $x = \pm\dfrac{\sqrt{2}}{2}$

into original function: $y = \left(\pm\dfrac{\sqrt{2}}{2}\right)^4 - 3\left(\pm\dfrac{\sqrt{2}}{2}\right)^2 + 2 = \dfrac{1}{4} - \dfrac{3}{2} + 2 = \dfrac{3}{4}$

$\left(\pm\dfrac{\sqrt{2}}{2}, \dfrac{3}{4}\right)$. Absolute minimum value of y and relative maximum

points are found using the sign analysis chart for y':

$$\begin{array}{ccccccc}
y' & - & 0 & + & 0 & - & 0 & + \\
\hline
x & \circled{-2} & -\frac{\sqrt{3}}{2} & \circled{-1} & 0 & \circled{1} & \frac{\sqrt{3}}{2} & \circled{2}
\end{array}$$

Candidate for maximum: $x = 0$ because y' changes sign from positive to negative. Substitute $x = 0$ into the original equation to find y : $y = 0^4 - 3(0)^2 + 2 = 2$. Relative maximum point: $(0, 2)$.

The absolute minimum occurs at $x = \pm\sqrt{\dfrac{3}{2}}$ because y' changes

sign from negative to positive. The y-value for these x-values

is $y = -\dfrac{1}{4}$.

2. Make a sign analysis chart for $f'(x)$ and $f''(x)$ based on the given graph.

$$\begin{array}{ccc}
y' & - & + \\
\hline
x & & -2
\end{array} \qquad \begin{array}{cc}
y'' & + \\
\hline
x &
\end{array}$$

original function is decreasing on $(-\infty, -2)$ and increasing on $(-2, +\infty)$ and it's concave up.

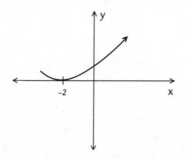

3. $v(0) = \langle -2, 0 \rangle$ $\qquad$ $|v(0)| = 2$

$$v = \left\langle \frac{dx}{dt}, \frac{dy}{dt} \right\rangle = \langle -2e^{-2t}, 6t \rangle \quad v(0) = \langle -2, 0 \rangle$$

$$|v| = \sqrt{(-2)^2 + 0^2} = 2$$

CHAPTER 9
SOLUTIONS

1. The top of the ladder is slipping at a rate of .013 ft/sec.

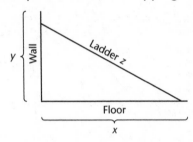

Let x = distance from the foot of the ladder to wall

y = distance from top of the ladder to floor.

Given: $z = 13$ ft, $\dfrac{dx}{dt} = 2$ in/sec, $\dfrac{dy}{dt} = ?$ when $x = 1$ ft

$$x^2 + y^2 = z^2 \rightarrow 2x\frac{dx}{dt} + 2y\frac{dy}{dt} = 0 \rightarrow 2(1)\left(\frac{1}{6}\right) + 2(\sqrt{168})\frac{dy}{dt}$$

$$= 0 \rightarrow \frac{dy}{dt} = -\frac{1}{3 \cdot 2\sqrt{168}} = -.013 \text{ ft/sec}$$

Notice that $\dfrac{dx}{dt} = 2$ in/sec was converted to $\dfrac{dx}{dt} = \dfrac{1}{6}$ ft/sec so that all units are in feet. Also, when $x = 12$ in = 1 ft, using the Pythagorean Theorem $y = \sqrt{168}$ ft.

2. $r = \sqrt{\dfrac{50}{3}}$

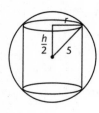

$$V_{cyl} = \pi r^2 h = \pi r^2 (2\sqrt{25 - r^2}) = 2\pi r^2 \sqrt{25 - r^2}$$

$$V'_{cyl} = 2\pi \left[r^2 \cdot \frac{-2r}{2\sqrt{25 - r^2}} + 2r\sqrt{25 - r^2} \right]$$

$$= 2\pi \left[\frac{-r^3 + 2r(25 - r^2)}{\sqrt{25 - r^2}} \right]$$

$$V'_{cyl} = 2\pi \left[\frac{-3r^3 + 50r}{\sqrt{25 - r^2}} \right] = 2\pi r \left[\frac{-3r^2 + 50}{\sqrt{25 - r^2}} \right] = 0 \rightarrow$$

$$r = 0 \text{ (reject)} \quad \text{or} \quad r = \sqrt{\frac{50}{3}}$$

Aside:

$$\frac{h^2}{4} + r^2 = 25$$

$$h^2 = (25 - r^2)4$$

$$h = 2\sqrt{25 - r^2}$$

Since V'_{cyl} changes from positive to negative at $r = \sqrt{\dfrac{50}{3}}$, V_{cyl} is a maximum at $r = \sqrt{\dfrac{50}{3}}$.

3. $\dfrac{dp}{dt} = \dfrac{3}{160}$

$$p\frac{dx}{dt} + x\frac{dp}{dt} - 10\frac{dp}{dt} = 3\frac{dx}{dt} \rightarrow \frac{13}{4}(-3) + 50\left(\frac{dp}{dt}\right) - 10\left(\frac{dp}{dt}\right)$$

$$= 3(-3) \rightarrow \frac{dp}{dt} = \frac{3}{160}$$

CHAPTER 10
SOLUTIONS

1. $c = \sqrt{3} + 1$

$$f'(c) = \frac{f(b) - f(a)}{b - a} \rightarrow -\frac{1}{(c-1)^2} = \frac{f(4) - f(2)}{4 - 2} \rightarrow -\frac{1}{(c-1)^2} =$$

$$\frac{-\frac{2}{3}}{2} \rightarrow -\frac{1}{(c-1)^2} = -\frac{1}{3} \rightarrow c = \pm\sqrt{3} + 1$$

$c = \sqrt{3} + 1$. Reject $c = -\sqrt{3} + 1$ since it is not in the given interval.

2. $c = \pm\dfrac{\pi}{3}, \pm\dfrac{2\pi}{3}, 0$

$f'(c) = 0 \rightarrow -6\sin(3c) = 0 \rightarrow \sin(3c) = 0 \rightarrow 3c = \pm\pi \rightarrow c = \pm\dfrac{\pi}{3}$.

Also, $3c = \pm 2\pi \rightarrow c = \pm\dfrac{2\pi}{3}$. And $3c = 0 \rightarrow c = 0$. Notice that

$3c = \pm 3\pi \rightarrow c = \pm\pi$ but these c values are the endpoints of the given interval and thus they must be rejected.

3. Yes, $c = e - 1$
   $y = \ln(x)$ does satisfy the Mean Value Theorem on $[1, e]$
   because it is continuous on $[1, e]$ and differentiable on $(1, e)$.

$$f'(c) = \frac{f(b) - f(a)}{b - a} \rightarrow \frac{1}{c} = \frac{f(e) - f(1)}{e - 1} \rightarrow \frac{1}{c} = \frac{1}{e - 1} \rightarrow c = e - 1.$$

4. No, $y = \ln(x)$ does not satisfy Rolle's Theorem on any interval
   because there are no x-values a and b such that $f(a) = f(b)$.

5. Because f is a polynomial, it continues and must obey the
   intermediate value theorem. There must be a root between
   $x = 1$ and $x = 2$. There is a root at $x = 3$ and there must be a root
   between $x = 4$ and $x = 5$.

CHAPTER 11
SOLUTIONS

1. $x \approx .6351203653$ $x = .6$ ERROR $= .0351203653$

   Enter $y = (5x - 3)^3$ in y_1 and n deriv(y_1, x, x) in y_2. Use the formula $x_{n+1} = x_n - \dfrac{f(x_n)}{f'(x_n)}$ and $x = 1$ (any number in $[0, 1]$ would do as a starting point)

   $x_1 = 1 - \dfrac{y_1(1)}{y_2(1)} = .8666669444$. Let Ans $= .8666669444$.

   $x_2 = \text{Ans} - \dfrac{Y_1(\text{Ans})}{Y_2(\text{Ans})} = .7777783796$

   $x_3 = $ Just press ENTER $x_3 = .7185195447$

   $x_4 = $ Just press ENTER $x_4 = .6790139673$

   $x_5 = $ Press ENTER $x_5 = .6526773843$

   $x_6 = $ Press ENTER $x_6 = .6351203653$

   The root of $y = (5x - 3)^3$ is $x = \dfrac{3}{5} = .6$
   ERROR $= |.6351203653 - 0.6| = .0351203653$

2. 19 iterations $x \approx .600$

3. $f(3.3) \approx .2321322258$

   $y_{n+1} = y_n + y'(x_n)\Delta x \rightarrow y_1 = \dfrac{1}{4} + \left(-\dfrac{1}{4^2}\right)(.1) = .24375 \rightarrow (3.1, .24375)$

   $y_2 = .24375 + \left(-\dfrac{1}{4.1^2}\right)(.1) = .23780116 \rightarrow (3.2, .23780116)$

   $y_3 = .23780116 + \left(-\dfrac{1}{4.2^2}\right)(.1) = .2321322258$

CHAPTER 12
SOLUTIONS

1. $A = 2$

$$A = \int_0^2 \left|1 - x^2\right| dx$$

$$A = \int_0^1 (1 - x^2)\, dx - \int_1^2 (1 - x^2)\, dx$$

$$A = x - \frac{x^3}{3}\Bigg]_0^1 - \left(x - \frac{x^3}{3}\right)\Bigg]_1^2$$

$$A = \left(1 - \frac{1}{3}\right) - \left[\left(2 - \frac{8}{3}\right) - \left(1 - \frac{1}{3}\right)\right] = \frac{2}{3} - 1 + \frac{7}{3} = 2$$

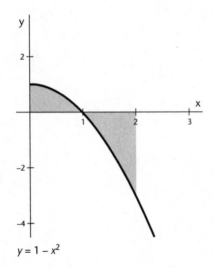

$y = 1 - x^2$

2. (A) 21 (B) –30

(A) $\int_a^b [3f(x) - 2g(x)]\, dx = 3\int_a^b f(x)\, dx - 2\int_a^b g(x)\, dx = 3(5) - 2(-3) = 21$

(B) $\int_b^a 6f(x)\, dx + \int_a^a \frac{g(x)}{\pi}\, dx = -\int_a^b 6f(x)\, dx + 0 = -6(5) = -30$

3. (A) $\sqrt{3x+1}$ (B) $\dfrac{8x}{e^{4x^2}}$

(A) $\dfrac{d}{dx}\displaystyle\int_0^x \sqrt{3t+1}\, dt = \sqrt{3x+1}$

(B) $\dfrac{d}{dx}\displaystyle\int_0^{4x^2} \dfrac{1}{e^t}\, dt = \dfrac{8x}{e^{4x^2}}$

4. 0

Average value of $y = \dfrac{1}{1-(-1)}\displaystyle\int_{-1}^{1}(ex^3 + x)\,dx = \dfrac{1}{2}\left(\dfrac{ex^4}{4} + \dfrac{x^2}{2}\right)\Bigg]_{-1}^{1} =$

$\dfrac{1}{2}\left[\left(\dfrac{e}{4} + \dfrac{1}{2}\right) - \left(\dfrac{e}{4} + \dfrac{1}{2}\right)\right] = 0.$

Notice that since $y = ex^3 + x$ is an odd function, its antiderivative on $[-a, a]$ is 0.

5. $\dfrac{1}{e^4}$

$\displaystyle\int_4^{\infty} e^{-x}\,dx = -e^{-x}\Bigg]_4^{\infty}\ \lim_{\ell\to\infty}(-e^{-\ell} + e^{-4}) = e^{-4} = \dfrac{1}{e^4}$

CHAPTER 13
SOLUTIONS

1. (A) 8.382332347   (B) 9.83182209

   (C) 9.143240618   (D) 9.107077219

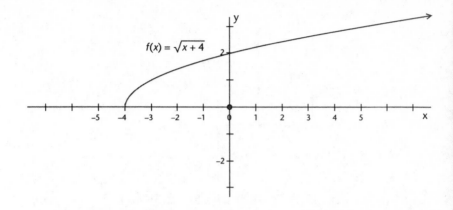

This is what the function should look like for parts (A) through (D) with their respective rectangles between $x = -3$ and $x = 2$.

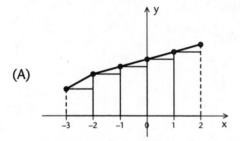

(A)

LRAM $f(x) = \sqrt{x+4}$

$A \approx 1(f(-3) + f(-2) + f(-1) + f(0) + f(1)) = 8.382332347$. This is an underestimation.

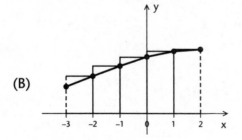

(B)

RRAM

$A \approx 1(f(2) + f(1) + f(0) + f(-1) + f(-2)) = 9.83182209$. This is an overestimation.

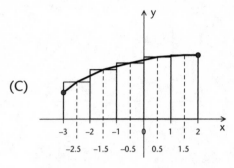

(C)

MRAM

$A \approx 1(f(-2.5) + f(-1.5) + f(-.5) + f(.5) + f(1.5)) =$
9.143240618.

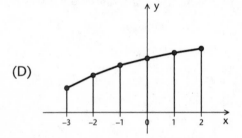

(D)

TRAP

$A = \dfrac{1}{2} \cdot 1(f(-3) + 2f(-2) + 2f(-1) + 2f(0) + 2f(1) + f(2)] =$

9.107077219.

CHAPTER 14
SOLUTIONS

1. $A = \pi$

   $A = \dfrac{1}{2} \int_{0}^{\pi} (2 \sin \theta)^2 \, d\theta = 3.141592654 = \pi.$ Using the area of a circle

   formula, $A = \pi r^2 = \pi(1)^2 = \pi.$

2. A = .8584073464

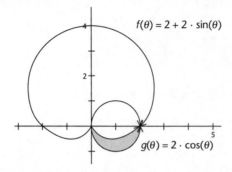

The area inside the circle and outside the cardioid:

$$A = \left(\frac{1}{2} \text{ CIRCLE}\right) - (\text{PIECE OF CARDIOID IN 4}^{\text{TH}} \text{ QUADRANT})$$

$$A = \frac{1}{2}(\pi) - \frac{1}{2} \int_{-\frac{\pi}{2}}^{0} (2 + 2\sin\theta)^2 \, d\theta = .8584073464$$

3. L = 2.662539877

$$\text{Curve length } = \int_a^b \sqrt{(x'(t))^2 + (y'(t))^2} \, dt = \int_0^1 \sqrt{e^{2t} + 4} \, dt =$$

2.662539877

4. V = 9.8696

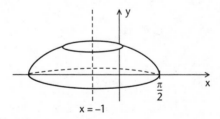

Cylindrical shell method $V = 2\pi \int_0^{\frac{\pi}{2}} (x+1)\cos x \, dx = 9.8696$

Washer Method $V = \pi \int_0^1 \{[\cos^{-1}(y)+1]^2 - 1^2\}dy = 9.8696$

5. $y = 2e^{x^2} - 1$

$\dfrac{dy}{dx} = 2x(y+1) \rightarrow \dfrac{dy}{y+1} = 2x\,dx \rightarrow \ln|y+1| = x^2 + c \xrightarrow{(0,1)} \ln 2 =$

$c \rightarrow \ln|y+1| = x^2 + \ln 2 \rightarrow \ln\dfrac{|y+1|}{2} = x^2 \rightarrow \dfrac{|y+1|}{2} = e^{x^2} \rightarrow |y+1| =$

$2e^{x^2} \rightarrow y+1 = 2e^{x^2}$ OR $y + 1 = 2e^{x^2}$. Since (0,1) satisfies $y + 1 = 2e^x$,
$y = 2e^{x^2} - 1$.

6. (A) When $P = 5$ (population is 5 million)   (B) 10 million

(A) Logistic growth equation $\dfrac{dP}{dt} = \dfrac{K}{M}P(M-P)$. Population grows

fastest when $P = \dfrac{M}{2}$. In $\dfrac{dP}{dt} = \dfrac{3}{10}P(10-P)$, $P = \dfrac{10}{2} \rightarrow P = 5$.

(B) $\lim\limits_{t\to\infty} P(t) = \lim\limits_{t\to\infty}\dfrac{M}{1+Ae^{-kt}} = M$. In this problem, $M = 10$. This means

that as time goes on, the population approaches 10 million.

CHAPTER 15
SOLUTIONS

1. $\dfrac{2}{3}e^{\frac{3x}{2}} + C$ u-substitution

$\int e^x \sqrt{e^x}\,dx \quad u = e^x \quad du = e^x\,dx$

$\int \sqrt{u}\,du = \int u^{\frac{1}{2}}\,du = \dfrac{2}{3}u^{\frac{3}{2}} + C = \dfrac{2}{3}e^{\frac{3x}{2}} + C$

2. $\dfrac{\sin^4(4x)}{16} + C$ *u*-substitution

$$\int \sin^3(4x)\cos(4x)\,dx \quad u = \sin(4x)$$

$$du = 4\cos(4x)\,dx \quad \frac{1}{4}du = \cos(4x)\,dx$$

$$\frac{1}{4}\int u^3\,du = \frac{1}{4}\frac{u^4}{4} + C = \frac{u^4}{16} + C = \frac{\sin^4(4x)}{16} + C$$

3. $\dfrac{x^2\ln x}{2} - \dfrac{x^2}{4} + C$ Integration by parts

$$\int x\ln x\,dx \quad u = \ln x \quad dv = x\,dx \quad \frac{x^2}{2}\ln x - \frac{1}{2}\int x\,dx = \frac{x^2}{2}\ln x - \frac{x^2}{4} + C$$

$$du = \frac{1}{x}dx \quad v = \frac{x^2}{2}$$

4. $4\ln|x-2| - 2\ln|x-1| + C$ Integration by partial fractions

$$\int \frac{2x}{x^2-3x+2}\,dx = \int \frac{2x}{(x-2)(x-1)}\,dx \quad \frac{A}{x-2} + \frac{B}{x-1} =$$

$$\frac{2x}{(x-2)(x-1)} \rightarrow Ax - A + Bx - 2B = 2x \rightarrow A + B = 2 \quad \text{and}$$

$$-A - 2B = 0 \xrightarrow[\text{EQUATIONS}]{\text{ADD}} -B = 2 \rightarrow B = -2 \rightarrow A = 4$$

$$\int \frac{2x}{x^2-3x+2}\,dx = \int \frac{4}{x-2}\,dx + \int \frac{-2}{x-1}\,dx = 4\ln|x-2| - 2\ln|x-1| + C$$

5. $\dfrac{2\sqrt{3}}{3}\tan^{-1}\left(\dfrac{x}{\sqrt{3}}\right) + C$ *u* = substitution

$$\int \frac{2}{3+x^2}\,dx = \frac{2}{3}\int \frac{1}{1+\frac{x^2}{3}}\,dx = \frac{2}{3}\int \frac{1}{1+\left(\frac{x}{\sqrt{3}}\right)^2}\,dx \quad u = \frac{x}{\sqrt{3}} \quad du = \frac{1}{\sqrt{3}}dx$$

$$\sqrt{3}\,du = dx$$

$$\frac{2\sqrt{3}}{3}\int\frac{1}{1+u^2}\,du = \frac{2\sqrt{3}}{3}(\tan^{-1}u)+C = \frac{2\sqrt{3}}{3}\tan^{-1}\left(\frac{x}{\sqrt{3}}\right)+C$$

CHAPTER 16
SOLUTIONS

1. 0

$$\lim_{n\to\infty}\left\{\frac{2n-1}{3en^2-1}\right\} = \lim_{n\to\infty}\left\{\frac{2n}{3en^2}\right\} = 0$$

2. (A) converges   Comparison test

    (B) diverges   Divergence test

    (C) converges   Ratio test for absolute convergence

(A) $\displaystyle\sum_{k=1}^{\infty}\frac{\cos k}{k^2}$ $\displaystyle\sum_{k=1}^{\infty}\left|\frac{\cos k}{k^2}\right| = \sum_{k=1}^{\infty}\frac{|\cos k|}{k^2} \le \sum_{k=1}^{\infty}\frac{1}{k^2}$ convergent

p series. By the comparison test $\displaystyle\sum_{k=1}^{\infty}\left|\frac{\cos k}{k^2}\right|$ converges

therefore $\displaystyle\sum_{k=1}^{\infty}\frac{\cos k}{k^2}$ converges. Remember, if a series

converges absolutely, then it converges.

(B) $\displaystyle\sum_{k=1}^{\infty}\frac{e^k}{k+1}$. Since $\displaystyle\lim_{k\to\infty}\frac{e^k}{k+1}\neq 0$, series diverges.

(C) Converges   Ratio test for absolute convergence

$$\sum_{k=1}^{\infty}\frac{(-1)^k}{(k-1)!} \qquad \lim_{k\to\infty}\left|\frac{(-1)^{k+1}}{(k+1-1)!}\cdot\frac{(k-1)!}{(-1)^k}\right| = \lim_{k\to\infty}\frac{(k-1)!}{k!}$$

$$= \lim_{k\to\infty}\frac{1}{k} = 0$$

CHAPTER 17
SOLUTIONS

1. A Maclaurin polynomial is a Taylor polynomial centered at $x = 0$.

2. Interval of convergence $1 \leq x \leq 3$, radius of convergence: 1

$$\sum_{k=1}^{\infty} \frac{(2-x)^k}{k^3} \quad \lim_{k \to \infty} \left| \frac{(2-x)^{k+1}}{(k+1)^3} \cdot \frac{k^3}{(2-x)^k} \right| = \lim_{k \to \infty} \left| (2-x) \frac{k^3}{(K+1)^3} \right| =$$

$$\lim_{k \to \infty} |2-x| = |2-x| < 1 \to -1 < 2-x < 1 \to 1 < x < 3$$

Check endpoints! At $x = 1$ $\displaystyle\sum_{k=1}^{\infty} \frac{1}{k^3}$ convergent p series. At $x = 3$

$\displaystyle\sum_{k=1}^{\infty} \frac{(-1)^k}{k^3}$ converges absolutely so it converges. Interval of

convergence is $1 \leq x \leq 3$. Radius is 1.

3. $P_4(x) = 3e + 3e(x-1) + \dfrac{3e}{2}(x-1)^2 + \dfrac{3e}{6}(x-1)^3 + \dfrac{3e}{24}(x-1)^4$

$P_4(x) = f(a) + f'(a)(x-a) + \dfrac{f''(a)(x-a)^3}{2!} + \dfrac{f'''(a)(x-a)^3}{3!} +$

$\dfrac{f^{IV}(a)(x-a)^4}{4!}$

$a = 1$, $f(x) = 3e^x$ $\quad f'(x) = f''(x) = f'''(x) = f^{IV}(x)$
$f(a) = f(1) = 3e$
$f'(a) = f'(1) = 3e$
$f''(1) = f'''(1) = f^{IV}(1) = 3e$
$P_4(x) = 3e + 3e(x-1) + \dfrac{3e}{2}(x-1)^2 + \dfrac{3e}{6}(x-1)^3 + \dfrac{3e}{24}(x-1)^4$

$$= 2x - \frac{2^3 x^2}{3!} + \frac{2^5 x^4}{5!} - \frac{2^7 x^6}{7!} + \cdots - \infty < x < \infty$$

laurin series for $y = \sin x$ is

$$\sin x = x - \frac{x^3}{3!} + \frac{x^5}{5!} - \frac{x^7}{7!} \quad -\infty < x < \infty$$

$$\sin(2x) = 2x - \frac{(2x)^3}{3!} + \frac{(2x)^5}{5!} - \frac{(2x)^7}{7!} \quad -\infty < x < \infty$$

$$\frac{\sin 2x}{x} = 2 - \frac{1}{x}\frac{(2x)^3}{3!} + \frac{1}{x}\frac{(2x)^5}{5!} - \frac{1}{x}\frac{(2x)^7}{7!} \quad -\infty < x < \infty$$

$$\frac{\sin 2x}{x} = 2 - \frac{2^3 x^2}{3!} + \frac{2^5 x^4}{5!} - \frac{2^7 x^6}{7!} + \quad -\infty < x < \infty$$